# Marine Corrosion Book

Copyright © 2011
by
*Marine Technical Training*

## Marine Technical Training
### Academy

*By: Alvaro Lopez*
Edition 2022

**Marine Technical Training** Academy

**Marine Technical Training** is an online academy specialized in technical and vocational courses according to the USCG standards, ABYC standards, Federal Regulations, and NMEA standards for Marine Engineers, Naval Architects, and Marine Master Technicians.

This book is the recommended textbook material for the **Marine Corrosion** course **MTT 1062** of 80 Hours, or three College credits.

Scan the following code to Enroll in this course.

# About the Marine Engineering Program.

This is a technical program focused on the training of Marine Master Technicians to cover the disciplines involved in the service, reparation, design, and refitting of pleasure vessels.

The courses are blended courses, 60 % online (Through pre-recorded episodes of 45 min following the curriculum corresponding to the textbook of each course, divided into chapters), 20% in conferences, and 20 % in remotely assisted labs.

The author of the videos and books is Engineer **Alvaro Lopez** with 35 years of experience teaching in Mechanical Engineering faculties and Naval Institutes.

The material used in this program is the property of **Marine Technical Training**. It is exclusive material with its own copyrights.

This program is composed by the following courses:

| | |
|---|---|
| • -Intro to Marine Engineering | MTT 1004 |
| • -Marine Basic Electricity | MTT 1400 |
| • -Marine Advance Electricity | MTT 2420 |
| • -Marine Electronics | MTT 2490 |
| • -Marine Diesel Engines I | MTT 1040 |
| • -Marine Diesel Engines II | MTT 2041 |
| • -Marine Gasoline Engines | MTT 1073 |
| • -Marine Transmissions | MTT 2234 |
| • -Marine Generators | MTT 2042 |
| • -Marine Auxiliary Systems | MTT 2541 |
| • -Marine Air Conditioning Systems | MTT 1543 |
| • -Marine Corrosion Systems | MTT 1062 |
| • -Marine Composite Materials | MTT 1312 |

# Copyrights ©

# ABYC Corrosion Certification

This book was designed like a consult handbook for marine technicians, naval engineers, shipyard managers, and boat owners interested to learn how the oxidation process can affect its vessels. in this book, you will find the procedure to do the wiring in a proper way to avoid future corrosive issues, and also you will find techniques to identify current leaks and how to solve it. In the content you will find QR codes to get access to some course video clips, wiring diagrams and posters specialized in each topic Also, it can be used as a complementary study guide to take the **ABYC Corrosion Certification.**

The cost of the ABYC certification is NOT part of this Book / Course.

# Table of Contents

# Table of Contents

# CHAPTER
# 1

## Molecular And Chemical Theory

### TOPICS

### Video Chapter 1 EP 1: Molecular And Chemical Theory

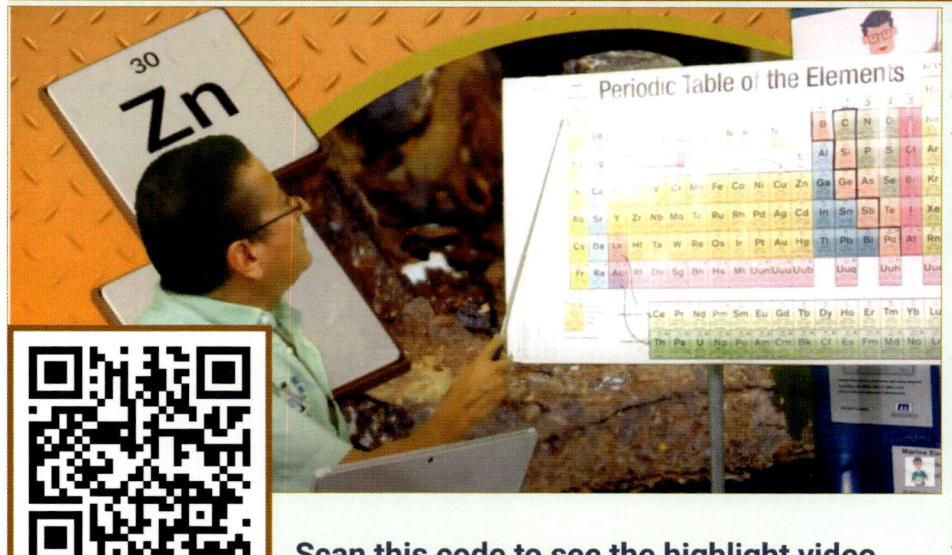

In this video, you will learn the fundamentals of molecular theory to understand how is the interaction between Oxigen with Metals and Non-metals. In this episode, we will study why some metals are not appropriate for marine environments. Also, we will discover the properties of some special elements and how is the organization of those elements in the periodic table.

**Scan this code to see the highlight video**

**Follow me**

# Matter

The **matter** is everything around you.
Atoms and molecules are all composed of matter.
The matter is anything that has mass and takes up space.

# States of Matter

The matter is found in 3 major states; solid, liquid, and gas.

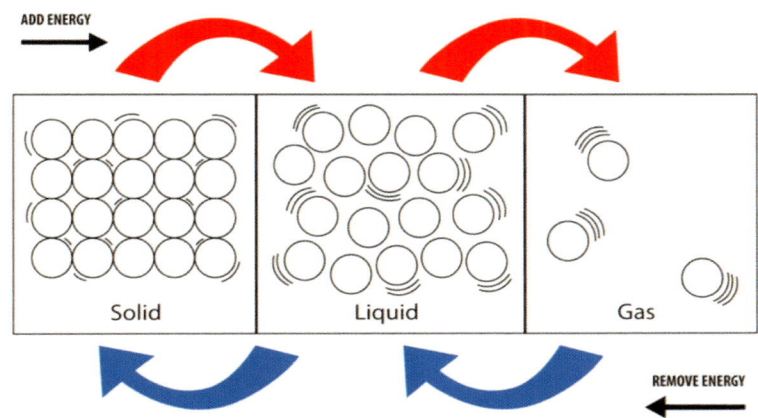

# The Nature of Matter

All matter is made of **atoms** composed of **P**rotons, **N**eutrons, and **E**lectrons.
The center, or nucleus, of the atom, is composed of positively charged protons and neutral neutrons. The outside of the atom has negatively charged electrons in various orbits.

Most atoms have the same number of Protons and Electrons, However, the number of Neutrons does not affect the identity of the element. The amount of Neutrons in its nucleus just affects the total weight.

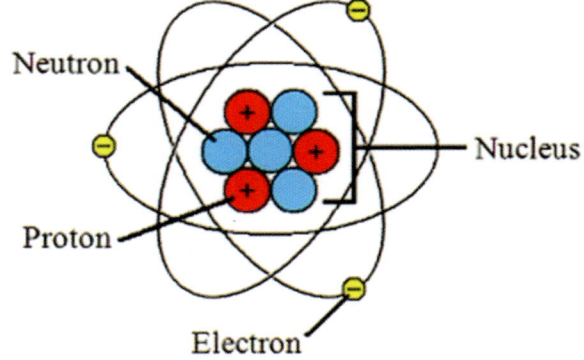

# Atomic Number

The Atomic Number of an element refers to the number of protons or electrons that make up an atom of that element.

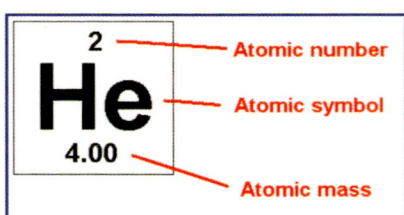

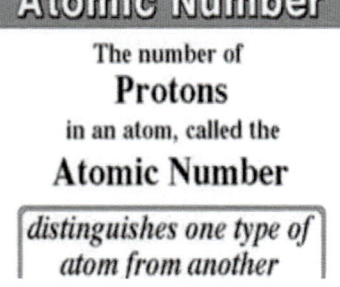

Atomic Number

The number of **Protons** in an atom, called the **Atomic Number**

*distinguishes one type of atom from another*

# Atomic Mass

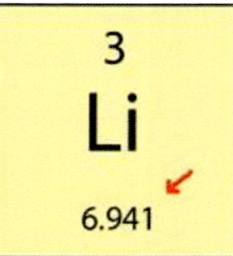

The Atomic mass (Weight) of the atom is determined by the number of protons and neutrons in the nucleus (the electrons are so small as to be almost weightless).
The proton has 1836 times the mass of the electron, but exactly the same size charge, only positive rather than negative.

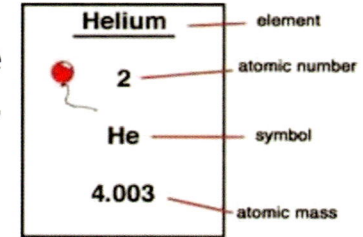

# Valence Electrons

Valence electrons are the electrons located in the outer orbit. They are the electrons that form bonds with other elements in compounds and that generally determine the properties of elements. Aluminum has three valence electrons because it is located in the main group number three. Oxygen and the rest of the elements located at group number six have six electrons in their outer orbit.

The elements with more electrons in the last orbit are considered elements with high levels of energy (Electronegativity). Those elements tend to attract electrons from other elements with lower levels of energy.

In the future, we will call Anodes elements that release electrons and Cathodes to elements that attract electrons.

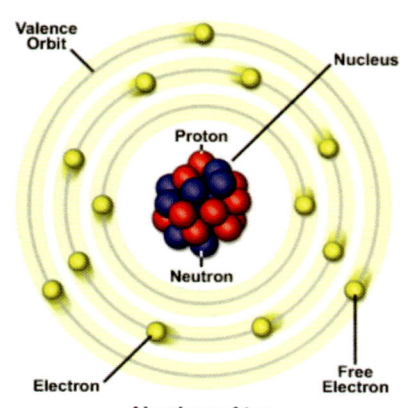

Mr. Lopez Classes.com
For Marine Engineers

# Electronegativity

Electronegativity is a measure of the tendency of an atom to attract a bonding pair of electrons. In other words, electronegativity is the amount of energy of each atom. Fluorine (the most electronegative element) is assigned a value of 4.0, and values range down to calcium and francium which are the least electronegative at 0.7.
The elements with more electronegativity tend to attract electrons from other elements with less energy.

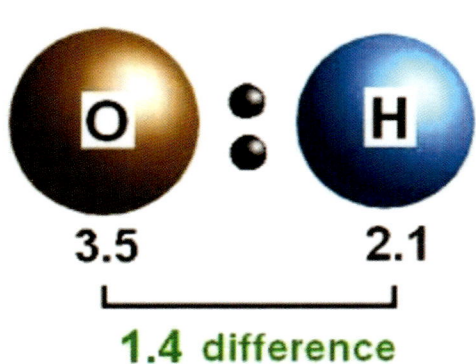

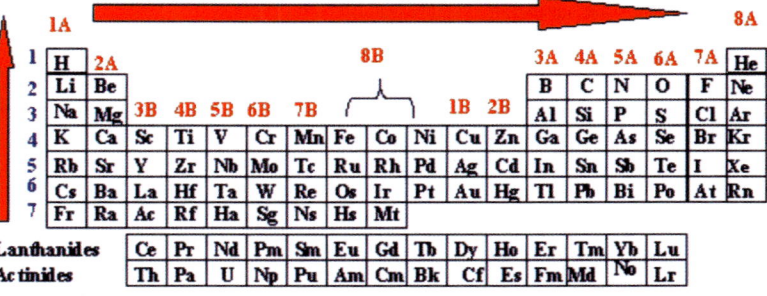

Electronegativity Trends in Periodic Table

Electronegativity increases from bottom to top in a column.
Electronegativity increases from left to right across a group.

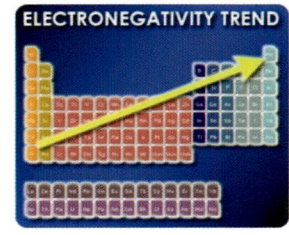

The higher the electronegativity, the greater the desire to gain electrons. Based on this term, elements belonging to groups 6 and 7 have high electronegativities as they seek additional electrons to complete their electron shells (8 electrons maximum in the outer orbit).

Elements in groups 1 and 2, this is elements with 1 or 2 electrons in their outer orbit have low electronegativities. those elements tend to release those electrons when they combine with other elements with higher levels of energy such as O, Fl, or Cl.

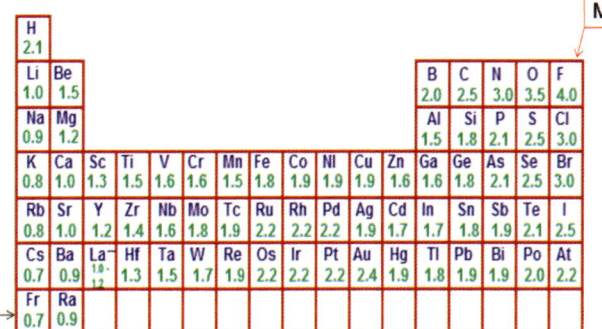

In general terms, this book is dedicated to studying how some metals and alloys are corroded by oxygen.
Due to their highest electronegativity (3.5), oxygen tries to attack the rest of the metals and non-metals producing oxides.

# What is the Octet Rule?

The octet rule is a rule that states that all atoms "want to have" **8 or 0** outer electrons called valence electrons.

In this example, Oxygen with more electronegativity attracts two electrons from hydrogen atoms to complete 8 electrons in the outer orbit.

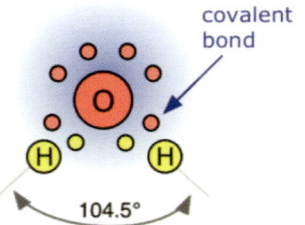

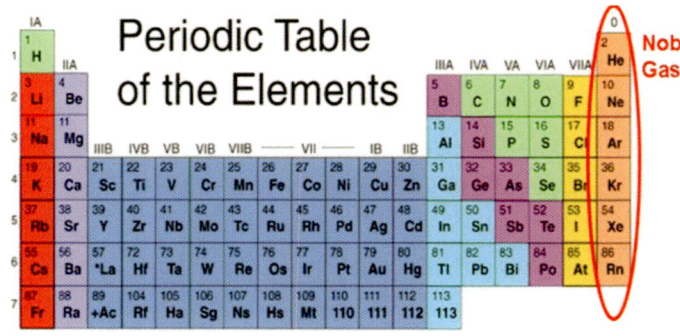

When the atom has eight or cero electrons in the outer orbit, the atom will be stable.

The elements located in group 8A of the periodic table are called 'Inert Gases" or "Noble Gases" because they have 8 electrons in the outer orbit. They don't react with any other element.

# Atom Properties

Atom size decreases as you move from left to right across the table and increases as you move down a column.

The energy required to remove an electron from an atom increases as you move from left to right and decreases as you move down a column.

The ability to form a chemical bond increases as you move from left to right and decreases as you move down a column.

# How the Periodic Table is Organized?

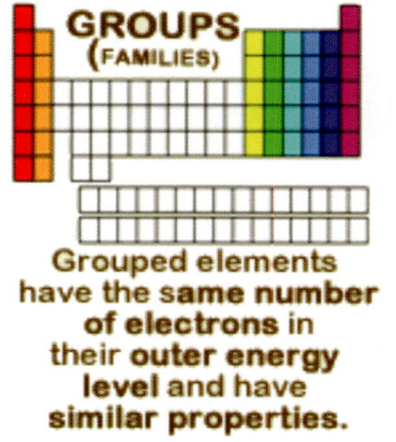

Grouped elements have the **same number of electrons** in their **outer energy level** and have **similar properties.**

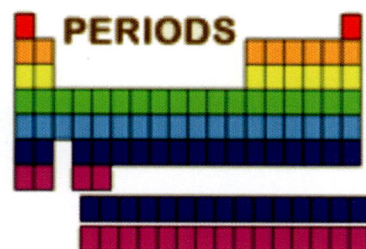

Each element is placed in a specific location because of its atomic structure and electronegativity; organized in groups (vertical columns) and periods (horizontal rows)

Today, the periodic table organizes 112 named elements and acknowledges several more unnamed ones.

# The Periods

When you look at the periodic table, each row is called a **period.**
All of the elements in a period have the same number of <u>atomic orbitals.</u>

All of the elements in the second row (the second period) have two orbitals for their electrons.
As you move down the table, every row adds an orbital. At this time, there is a maximum of seven electron orbitals.

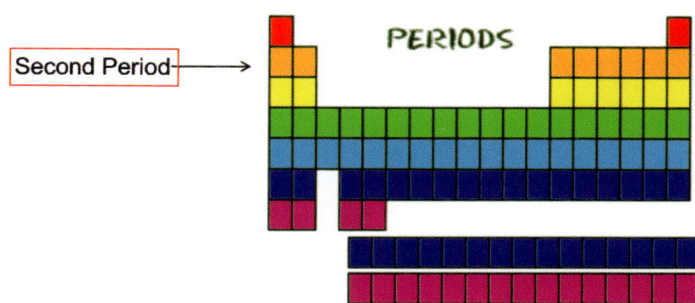

# The Groups

The elements are also organized in vertical columns, or **groups**, based on similar physical characteristics and chemical behavior. The elements in each group have the same number of electrons in the last valence shell.

Group 1A elements are the **alkali metals** and group 2A are the **alkaline earth**. Group 7A are the **halogens** and group 8A are the **noble gases.**

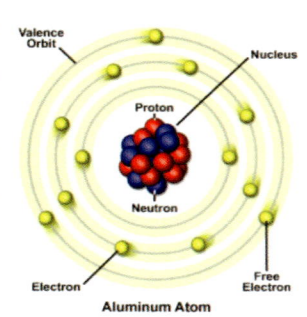

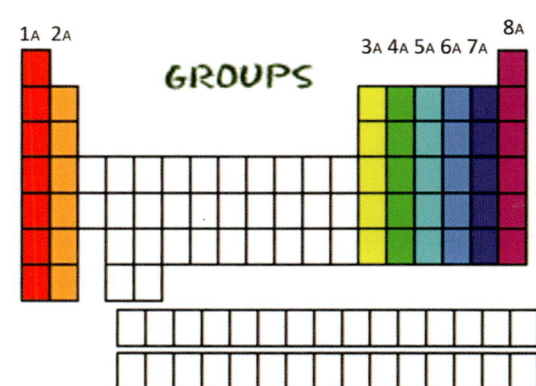

# Metals and Non-Metals

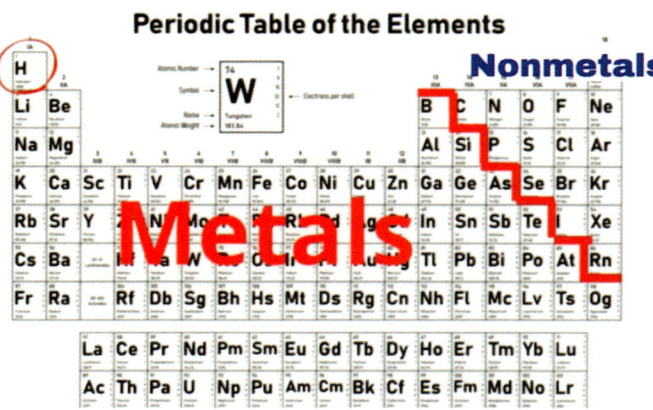

The elements located at the left side of the Red division are considered metals, otherwise, the elements located at the right side are called non-metals. Hydrogen is a nonmetal located at the right side based on its structure (one electron in its valence orbit) however, it is a nonmetal.

# Transition Metals Valence Electrons

The transition metals are an array of ten sub-groups (red rectangle) located in the middle of the periodic table. The elements in the first nine groups have an atomical structure with three (3) electrons in their outer orbit

The elements in the last group: zinc (Zn), cadmium (Cd), and mercury (Hg) have two (2) electrons in their outer orbit, however, all of them are considered transition metals for other atomical and mechanical properties.

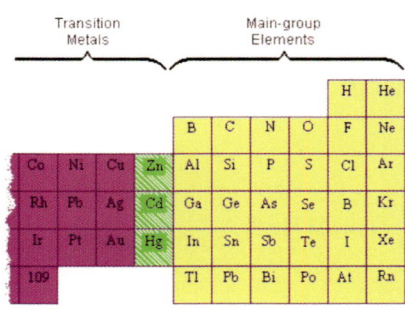

Oxigen located in group number six is one of the elements with a higher level of energy in the environment. When different metals including Zn are exposed to the environment, the oxygen prefers to attack the Zn because the oxidation process is too simple. Zinc donates the two outer electrons and Oxygen attracts those electrons to complete eight electrons. Now the molecule Zn O is a stable molecule.

The transition elements are located in between the main groups 2A and 3A. These define the 'B' groups. Since they are all metals, these elements are referred to as **transition metals.**

Transition metals are superior conductors of heat as well as electricity. They are malleable, which means they can be shaped into sheets, and ductile, which means they can be shaped into wires.

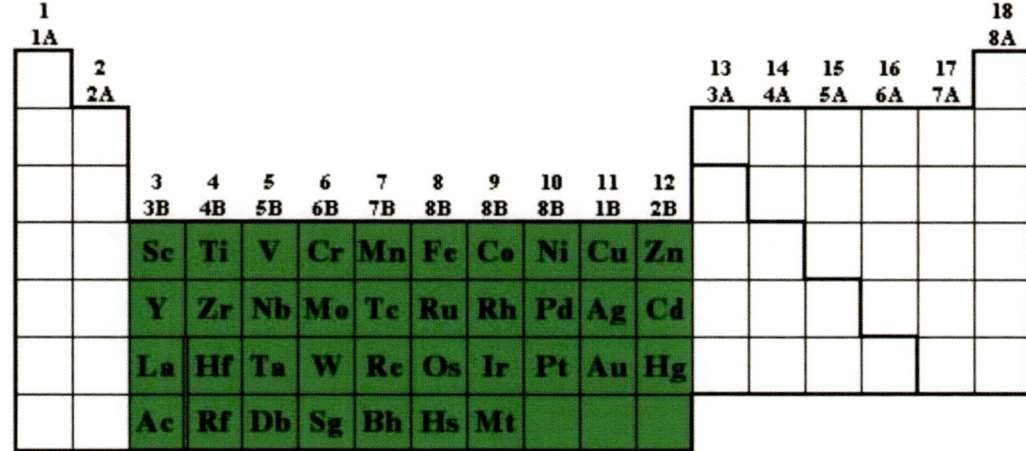

# Molecule and Compound

A molecule is formed when two or more <u>atoms</u> join together chemically.

A compound is a molecule that contains at least two <u>different</u> elements. All compounds are molecules but not all molecules are compounds.

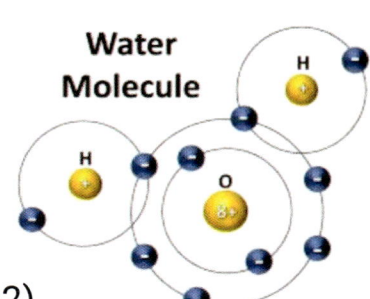
Water Molecule

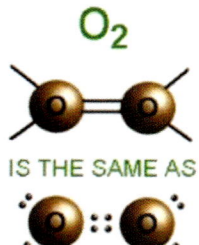

Molecular <u>hydrogen</u> ($H_2$), molecular <u>oxygen</u> ($O_2$), and molecular <u>nitrogen</u> ($N_2$) do not compound because each is composed of a single element.

Water ($H_2O$), carbon dioxide ($CO_2$), and methane ($CH_4$) compound because each is made from more than one element.

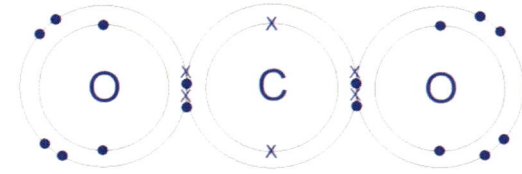

# The Water Molecule

In the water molecule, hydrogen can also be happy with no electrons due to the fact that there is no energy level below the first one, so zero electrons would also be a full outer shell. and Oxigen now with eight electrons in their outer shell will be stable too.

So water is a stable molecule that normally does not react. this type of bonding is permanent and strong. It is called an Ionic bond. Two different atoms with different levels of energy.

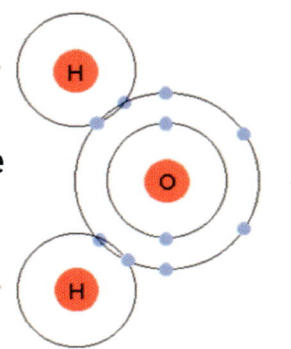

# The Oxygen molecule

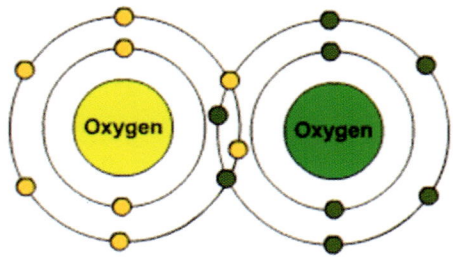

This molecule results from the covalent bonding of two oxygen atoms which share every two atoms of its outer orbit. In this way, the molecule is stable because both atoms have 8 electrons in the outer orbit.

This union is temporary for this reason oxygen layer is easily altered.

Marine Technical Training Academy

Mr. Lopez Classes.com
For Marine Engineers

# Why Electrons Move ?

**Excitation** is the process where the electron in an atom gains energy and due to its high energy it cannot reside in that particular orbit and it goes to the orbit with higher energy. This process is called Excitation.

The most common way for electrons to move to an excited state is by absorption of electromagnetic radiation.

They come back to the ground state at the first possible opportunity because in the excited state they have lots of potential energy.

This is why a covalent bond (two identical atoms sharing electrons) is considered non-permanent in comparison with an Ionic bonding or permanent.

# Chemical Bonding

A chemical bond is an attraction between atoms that allows the formation of chemical substances that contain two or more atoms.

Chemical compounds are formed by the joining of two or more atoms.

The electrons that participate in chemical bonds are the valence electrons.

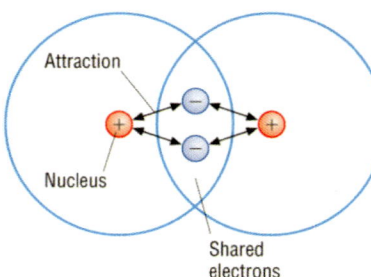

Electrons repel each other, yet they are attracted to the protons within atoms. The interplay of forces results in some atoms forming <u>bonds</u> with each other and sticking together.

## Two types of chemical bonds

| <u>Ionic bonds</u> occur . . . | <u>Covalent bonds</u> occur . . . |
|---|---|
| when anions and cations attract by virtue of their opposite charges | when atoms share valence electrons. |
| between metal and non metals | between two non metals |

# Ionic and Covalent Bonds

When two atoms share electrons, they form a **covalent bond**. When one atom takes an electron away from another and the resulting positive and negative ions are attracted to each other, those atoms have formed an **ionic bond**.

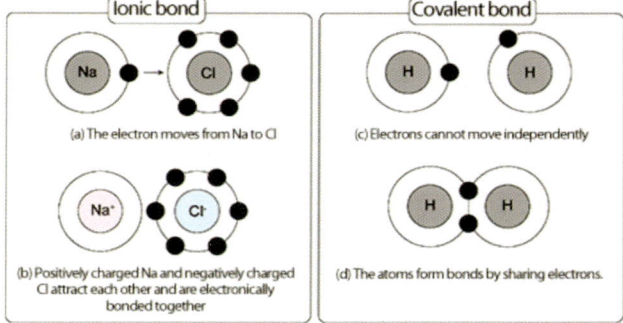

# Ionic Bond

An ionic bond is formed when one atom <u>accepts or donates</u> one or more of its valence electrons to another atom.

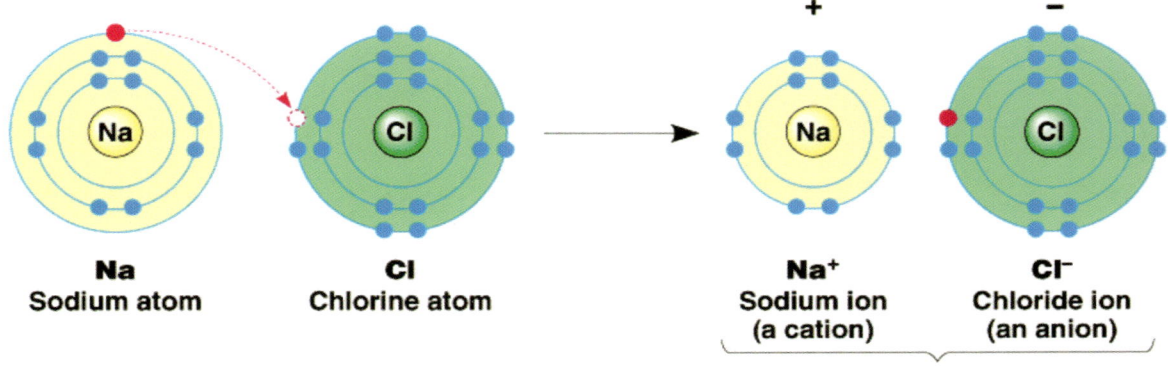

| Na | Cl | Na⁺ | Cl⁻ |
|---|---|---|---|

**Na**
Sodium atom

**Cl**
Chlorine atom

**Na⁺**
Sodium ion
(a cation)

**Cl⁻**
Chloride ion
(an anion)

Sodium chloride (NaCl)

# Covalent Bond

A **covalent bond** is formed when <u>atoms share</u> valence electrons.
The stable balance of attractive and repulsive forces between atoms when they share electrons is known as **covalent bonding.**

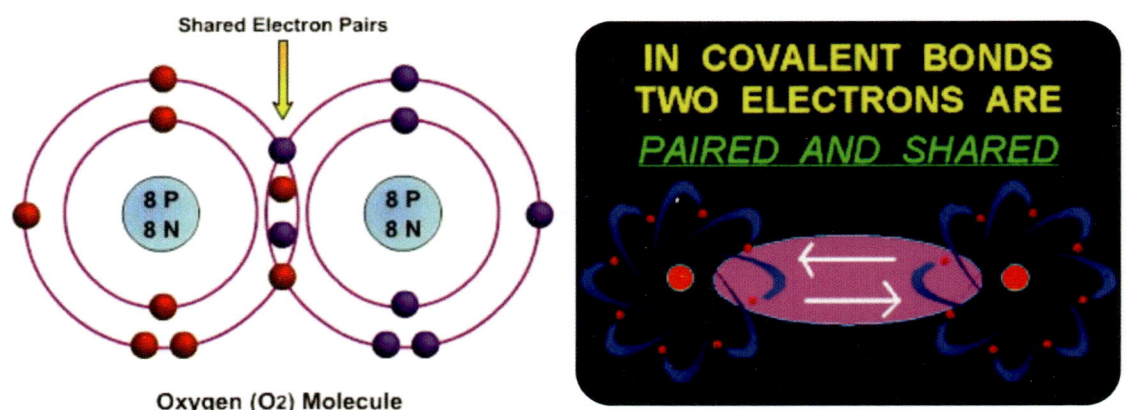

Oxygen (O₂) Molecule

IN COVALENT BONDS
TWO ELECTRONS ARE
*PAIRED AND SHARED*

### Video Chapter 1 EP 2: Ionization & Chemical Bonding

In this Video we are going to analyze how the ionization affect the bonding process in between molecules and in between metallic structures with the environment. Enjoy it

**Scan this code to see the highlight video**

**Follow me**

# Ionization

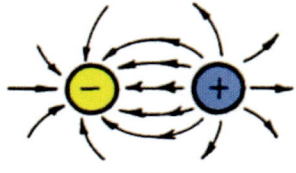

When energy is added to an atom by some exterior force such as heat, friction or bombardment by other electrons, the atom becomes excited.

# From Electron to Photon

A photon is produced whenever an electron in a higher-than-normal orbit falls back to its normal orbit.
During the fall from high energy to normal energy, the electron emits a photon -- a packet of energy -- with very specific characteristics.

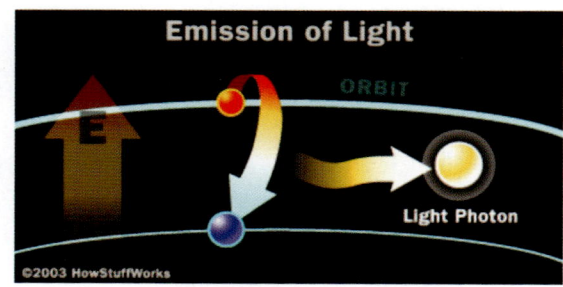

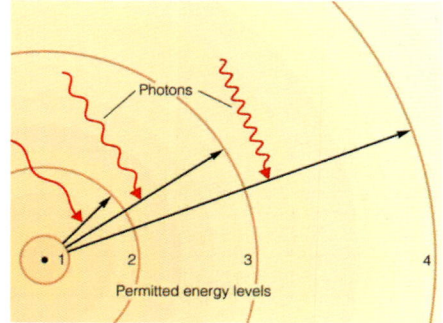

The photon has a frequency, or color, that exactly matches the distance the electron falls.
You can see this phenomenon quite clearly in gas-discharge lamps. Fluorescent lamps, neon signs, and sodium-vapor lamps.

# Ions

In the ground state, an atom will have an equal number of protons and electrons, and thus will have a net charge of 0. However, because electrons can be transferred from one atom to another, it is possible for atoms to become charged. Atoms in such a state are known as **ions**.

- If a neutral atom gains an electron, it becomes **negative**. This kind of ion is called an **anion**.
- If a neutral atom loses an electron, it becomes **positive**. This kind of ion is called a **cation.**

# Ionization

**Ions** are formed when atoms, or groups of atoms, <u>lose or gain</u> electrons.

Metals lose some of their electrons to form positively charged ions, e.g. $Fe+2$, $Al+3$, $Cu+2$, etc.

Nonmetals gain electrons and form negatively charged ions, e.g. $Cl-$, $O-2$, $S-2$, etc.

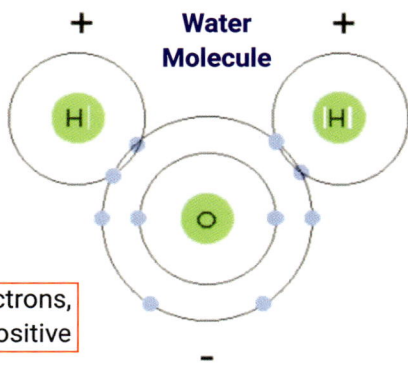

Hydrogen lose electrons, become positive

Oxygen gain electrons, become negative

# Anode (+) and Cathode (-)

An **Anode (+)** is an electrode that releases electrons when a flow of electricity passes through it. (Sacrificial Anode)

A **Cathode (-)** is an electrode that receives electrons when a flow of electricity passes through it.

The flow of current is opposite to the flow of electrons as I explained in my Electricity Handbook (Pag 19). The electrons migrate from the (+) terminal into the (-) terminal. They migrate from the **Anode** into the **Cathode**. However, the electrical current is flowing from the negative terminal (Cathode) into the positive terminal (Anode).

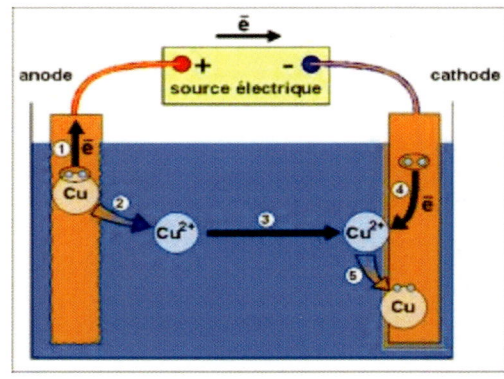

# Current Flow vs Electrons Movement

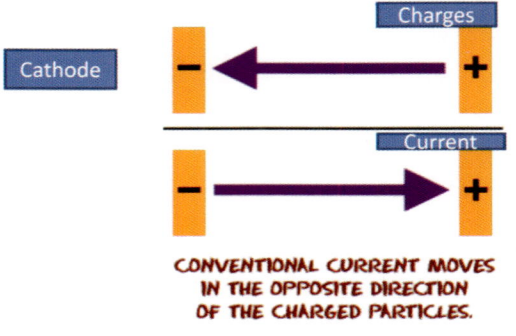

A good example is a lead-acid battery. After a year used the positive terminal will be fully eroded and covered with a green powder. It is eroded because each time an electron migrates a hole is produced at the surface metal. The green powder obeys the oxidation process of copper at the cable terminal (Cupric Oxide).

A similar situation occurs on the Sacrificial Anodes used in our boats. If the Anode is properly installed, it will be eroded gradually. As we will learn later in the Cathodic reaction the anode will be eroded each time that an electron migrates.

# Al and Zn Galvanic Cell

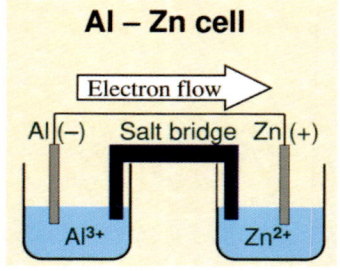

zinc connected to aluminum will form a corrosion cell, but in this case, the aluminum becomes the cathode and the zinc (anode) corrodes.
If this couple of metals are surrounded by salt water, then the oxygen will prefer attacks the Zn and

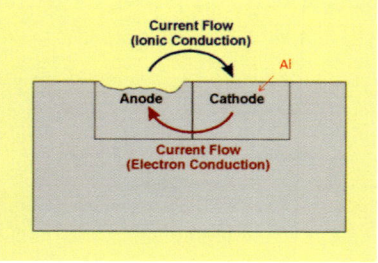

once again the Aluminum structure will be protected for the Anode of Zn.

# Outboards (Stainless & Aluminum)

In a marine environment, If your outboard is not protected with sacrificial anodes. The aluminum will work as an anode and the stainless parts as a cathode. Then the aluminum will be corroded due to the presence of the seawater (Electrolyte) and also by the flow of small currents coming from different sources such as the alternator, starter, ignition coil, etc.

## Copper & Aluminum

This is the worst metallic couple in a marine environment (Not accepted by ABYC). It is a combination of an aluminum structure connected with copper or a copper alloy (ie: bronze fittings). In this case, any small current moving through the metals accelerate the faster deterioration of the aluminum.

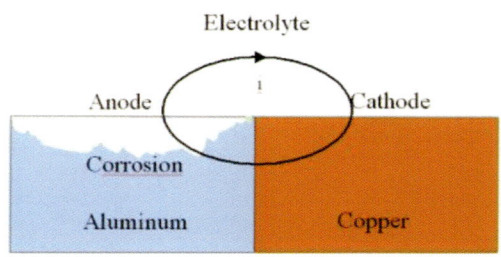

The copper with higher electronegativity (1.9) will be considered a more noble metal than Aluminum with (1.5). Then the more noble metal will attack the less noble producing erosion.

This process will be accelerated if a small current flows through the metals (Electrolytic Corrosion).

**Video Chapter 1 EP 3: Why Zinc is used such as sacrificial anode**

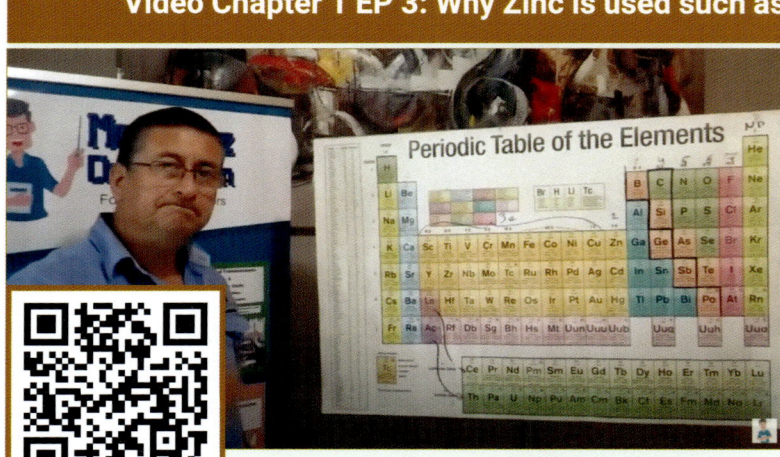

Scan this code to see the highlight video

This is a wonderful video to understand why Zinc is used such as a sacrificial anode to protect other metals from corrosion. In the video, Mr. Lopez explains in a simple way why "Zn" is an excellent sacrificial anode. Enjoy the video.

**Follow me**

## Reducing the Corrosion Process

There are five simple ways to reduce or stop the corrosion process in a boat:

First, selecting proper construction materials. Using exclusively marine-grade materials.

Second, draining constantly the static electricity and shorts to the ground by means of a good Ground System

Third, selecting the appropriate sacrificial anodes according to the hull material and type of water.

Four, checking periodically the quality of the boat bonding system using a Silver Chloride Electrode test method

Five. Installing an Amp-meter at the AC and DC panel to identify leaks of current.

# CHAPTER 2

## Metals and Naval Alloys

## TOPICS

## Video Chapter 2 EP 1: Metals and Non-Metals

In this video, you will learn how metals and non-metals are organized in the periodic table. Also, we are going to study the basic properties of each group and how oxygen is combined with both of them producing acidic oxides and basic oxides. Enjoy it

Scan this code to see the highlight video

Follow me

# Metals and Non-Metals

The elements in the periodic tables are classified into Metals, Non-metals, and Metalloids. Hydrogen is a nonmetal located at the right side based on its structure (one electron in its valence orbit) however, it is a nonmetal.

The combination of different metals produces Alloys and the alloys are divided into Ferrous alloys and Non-ferrous alloys.

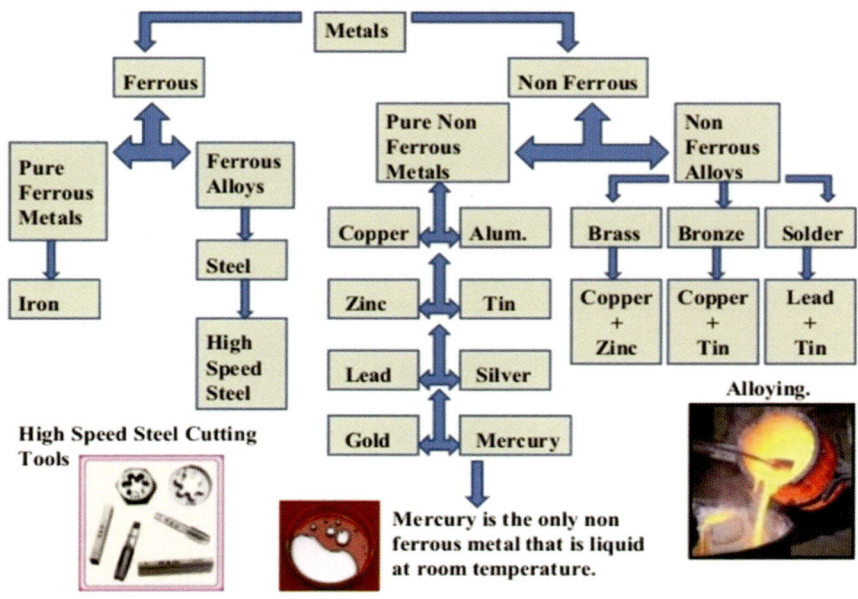

# Metals and Metal Alloys

Alloys are mixtures of two or more pure metals.

Alloys tend to have better strength properties than pure metals.

Alloys are classified as either: Ferrous or nonferrous.

Ferrous alloys contain iron as the main component.

In general ferrous alloys tend to corrode and therefore need some form of protection against corrosion.

Non-ferrous metals do not tend to corrode in the same way. However soon or later they will be corroded.

In both groups, there are Marine Grade Alloys. Those are special alloys with ultra-low or zero tendencies to be corroded.

# Marine Grade Alloys

To minimize the corrosion process the metallurgical engineers create some special Marine Grade Alloys.
There are Ferrous marine grade alloys and Non-ferrous marine grade alloys.

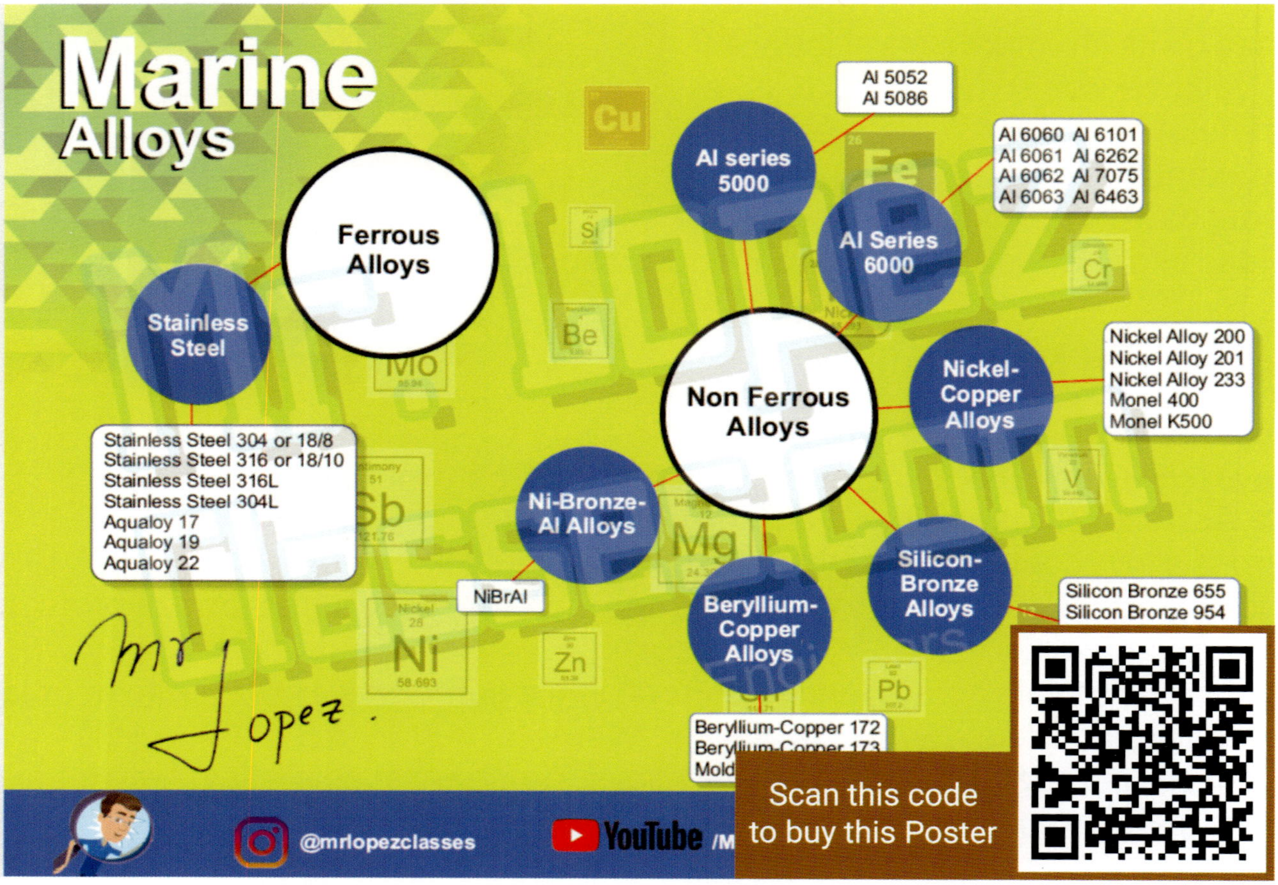

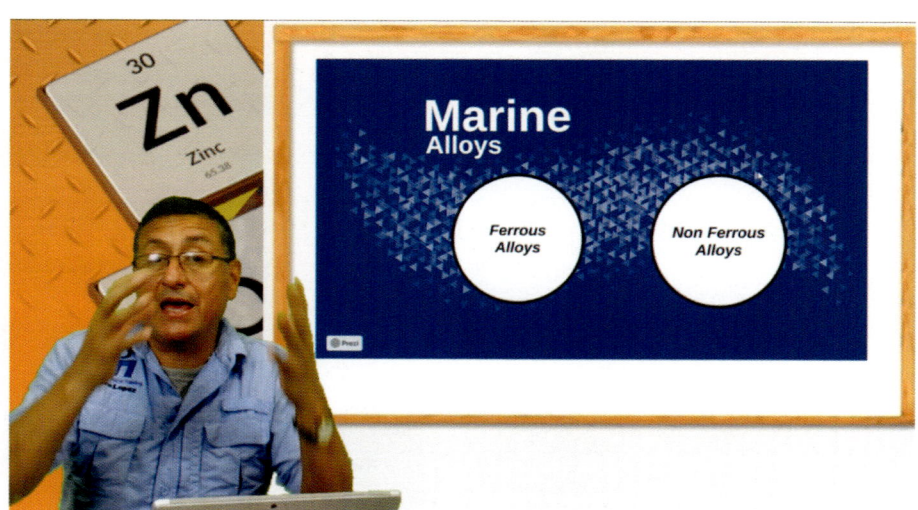

Scan this code
to watch the video

# Common Non-Ferrous Alloys

**Brass** = (45%-70%)Copper + Zinc (Non Marine Approved)
- More susceptible to corrosion "dezincification"
- Dezincification is when the zinc portion of the brass alloy is carried away from the mother metal as a soluble salt.

**Bronze** = 90% Copper + Tin (Propeller)

**Duralumin** = 4% Copper + 1% Mn + 95% Aluminum
- Their hardness can be increased by tempering (Aviation Uses)

**Solder** = Tin + Lead

**Monel 400 / 500** = 67% Nickel + 22% Copper + 1% Silica
- good corrosion resistance, good weldability, and high strength (Shafts)

**NiBrAl** = (3.5-4.5) Nickel + 79% opper + (8.5-9.5%) Aluminum + (0.8-1.5%) Mn
- High Resistance to Fatigue, Corrosion and wear (Propellers)

**Tobin Bronze** = 0.75% Tin + 39% Zinc + 59% Copper
- High strength, Toughness and Corrosion resistance (Propeller Shafts)

**Aluminum 6000 Series** 95% Al + Si + Mg
- Medium strength properties, easily anodized

**Aluminum 7000 Series** 97% Al + Zn + Si
- Excellent Fatigue properties

**MoldMax** = 97.5% Copper + 0.2% Ni + 0.1% Co + 2 % Be
- High Strength, Good thermal conductivity, Good wear resistance, Good Weldability

**Naval Bronze** = 85% Cu + 5% Tin + 5% Pb + 5% Zn
- High tensile Strength, Good thermal conductivity, Good for seawater contact

**Babbitt** = 86% Tin + 7% Cu + 7% Sn + Pb
- High heat resistance and Good Lubricity (Engine plain bearings)

# Common Marine Ferrous Alloys

**Stailess Steel 316 / 316L** = (16-18%) Cr + (10-14%) Ni + 0.08% C + 70% Fe
- Resists atmospheric corrosion in polluted marine environments

**Stailess Steel 304 / 304L** = (18%) Cr + (10%) Ni + 0.07% C + 2% Mn + 70% Fe
- Resists atmospheric corrosion in polluted marine environments

**Stailess Steel 18/8** = (18%) Cr + (8%) Ni + (Balance) Fe
- Used for food preparation and dining

**Stailess Steel 18/10** = (18%) Cr + (10%) Ni + (Balance) Fe
- Resists atmospheric corrosion in polluted marine environments

**Stailess Steel 18/0** = (18%) Cr + (0.75%) Ni + (Balance) Fe
- Reduced corrosion resistance. Has magnetic properties

**Aqualoy 17** = (15-17%) Cr + (3-5%) Ni + (1%) Mn + 0.07% C + Balance Fe
- Has the highest strength and hardness of all stainless steel for boats
- Provides the maximum corrosion resistance possible
- Used to fabricate propeller shafts

**Aqualoy 19** = (18-20%) Cr + (8-10.5%) Ni + (2%) Mn + 0.08% C + (0.2-0.3%) N + Fe
- Used to fabricate propeller shafts

**Aqualoy 22** = (22%) Cr + (12.5%) Ni + (5%) Mn + 0.25Si + 1% C + Fe
- Provides outstanding corrosion resistance
- Resists barnacle accumulation and marine organisms
- Used to fabricate propeller shafts

**Note: The letter "L" denotes Low content of Carbon**

# Pure Metals

Only Copper, Gold and Platinum, occur naturally in their elemental form.

Most metals occur in nature as oxides in ores, combined with some worthless junk like clay and silica.

Mercury is not usually found free in nature and is primarily obtained from the mineral cinnabar (HgS).

It is poisonous in soluble forms such as mercuric chloride or methylmercury.

# Ore

An **ore** is a type of rock that contains sufficient <u>minerals</u> with important elements including <u>metals</u> that can be economically extracted from the rock.

The ores must be processed to get the pure metals out of them.

# Ores Processing

The grade or concentration of an ore mineral, or metal, as well as its form of occurrence, will directly affect the costs associated with mining the ore.

The cost of extraction must thus be weighed against the metal value contained in the rock to determine what ore can be processed and what ore is of too low a grade to be worth mining.

## Metals and Water

When Oxygen attacks metals produce, basic oxides
Basic Oxide = Oxygen + Metal ($MgO$, $FeO$).
Basic Oxides + Water = Basic solution
$O\ Na + H_2O = Na\ (OH)_2$
React with acids to form salt & water.
Basic Oxides are usually insoluble in water. Those that dissolve in water form alkaline solutions.

## Alkali Metals

Sodium and Potassium are Alkali Metals. They are located in group I of the periodic table.

They are extremely reactive metals.
They react strongly with water to form strong alkalis.

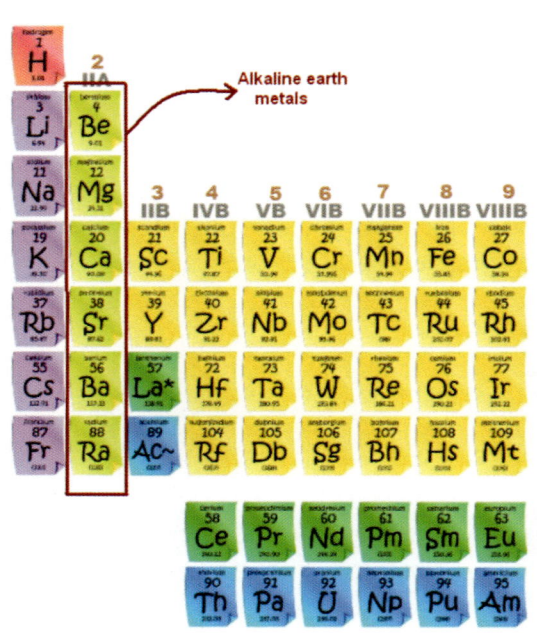

## Alkali Metals Properties

The alkali metals have the following properties in common:

- they have low melting and boiling points compared to most other metals.
- they are very soft and can be cut easily with a knife.
- they have low densities (lithium, sodium, and potassium will float on water).
- they react quickly with water, producing hydroxides and hydrogen gas.
- their hydroxides and oxides dissolve in water to form alkaline solutions.

# Alkali Metals, Halogens, Noble Gases

The alkali metals are soft, reactive metals, The halogens are reactive non-metals, The noble gases are unreactive non-metals, which exist as single atoms.

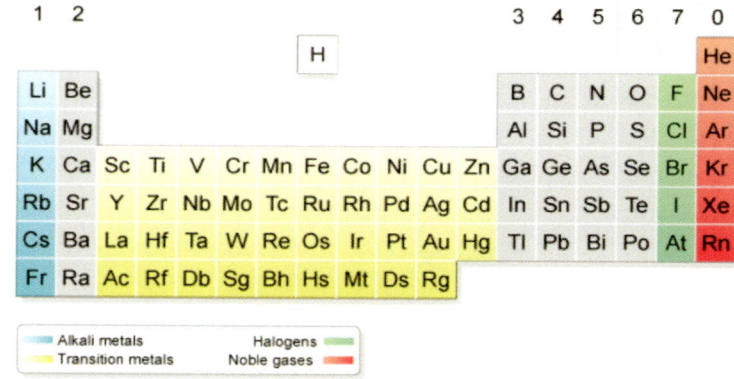

# Transition Metals

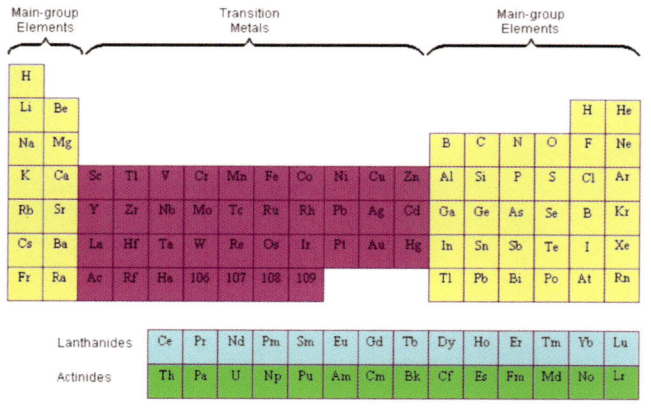

The elements in the periodic table are often divided into four categories: (1) main group elements, (2) transition metals, (3) lanthanides, and (4) actinides.

There is some controversy about the classification of the elements on the boundary between the main group and transition-metal elements on the right side of the table. The elements in question are zinc (Zn), cadmium (Cd), and mercury (Hg).

# Boundary Metals

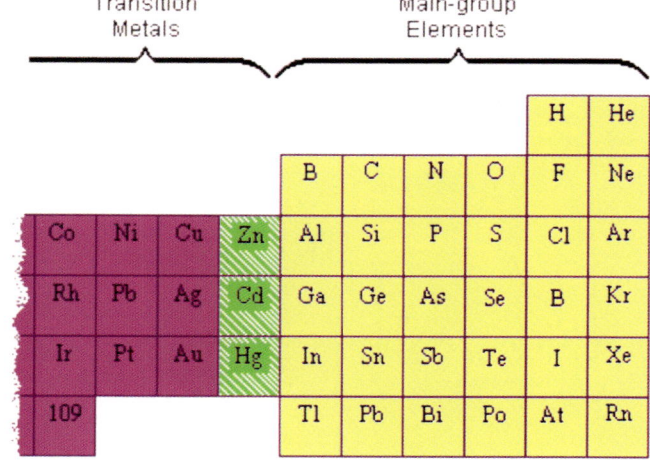

## Transition Metals vs. Main-Group Elements

**Zn** has two electrons in the valence orbital, for that reason, it is attractive for oxygen. The majority of the transition metals except for the group of Zn works with three electrons in the valence orbit.

## Sacrificial Anodes

The most common oxidation states of the boundary group ( Zn, Cd, Hg) occur with two electrons on the valence orbit.
The rest of the transition metals works with three electrons on the valence orbit.
That is the reason why zinc is used like protective metals ( Sacrificial anodes) in order to avoid corrosion with other metals.

## Prevention of Corrosion

Oxygen prefers to attack metals with only two electrons in its outer shell. In this way, the oxygen will have eight electrons and will be stable.
Zinc is a good example, for this reason, it is used to avoid corrosion of other metals.

Zinc Oxide ⟶ 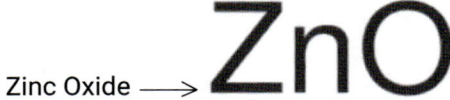 ZnO

## Extracting Metals
### Metallurgical process

# Extracting Gold

Some metals, such as gold, are found naturally as pure metals in rocks.

Gold is unreactive, so it does not combine with other elements.

To extract gold from its ore, huge grinders crush the ore to a fine powder.

The powder is mixed with a solution of cyanide. Only the gold from the ore dissolves in the solution.

Powdered zinc is added to bring the gold out of the solution. The gold is melted down and poured into molds.

# Aluminum Extracting

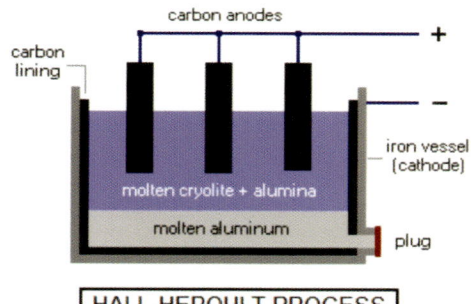

HALL-HEROULT PROCESS

The usual aluminum ore is bauxite. **Bauxite** is essentially an impure aluminum oxide. Aluminum is too high in the electrochemical series (Electronegativity) to extract from its ore using carbon reduction.

The ore is first converted into pure aluminum oxide by the Bayer Process, and this is then electrolyzed in solution in molten cryolite - another aluminum compound.

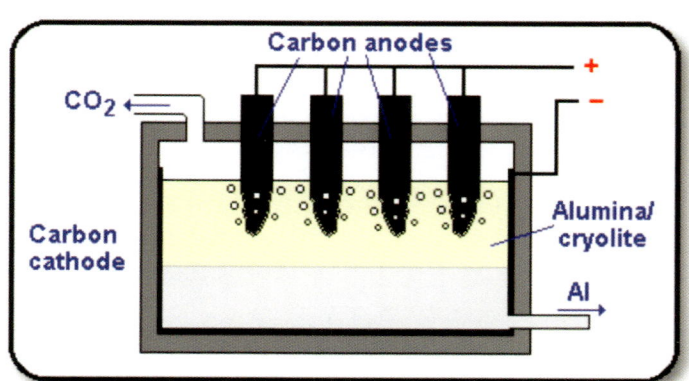

# The Bayer Process

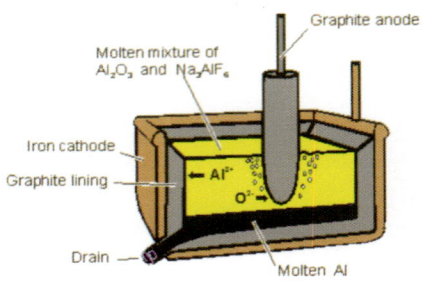

Crushed bauxite is treated with moderately concentrated sodium hydroxide solution Na(OH).

The impurities in the bauxite remain as solids.

All of these solids are separated from the sodium by filtration.

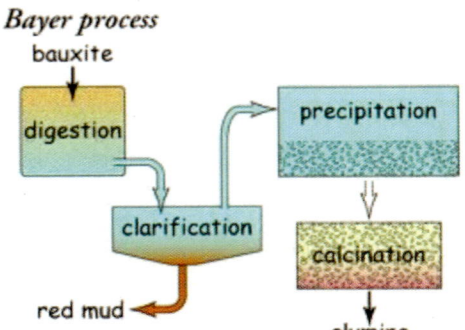

They form a "red mud" which is just separated from pure aluminum.

The aluminum is drained into molds to produce ingots.

# The Product

Around 6 tons of Bauxite and 2 tons of alumina are required to produce one ton of pure aluminum.

## Aluminum by Countries

Japan, Kazakhstan, and China produce more than 90% of the aluminum in the world.

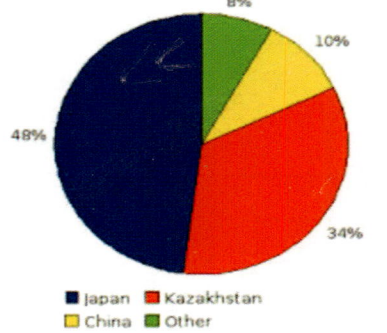

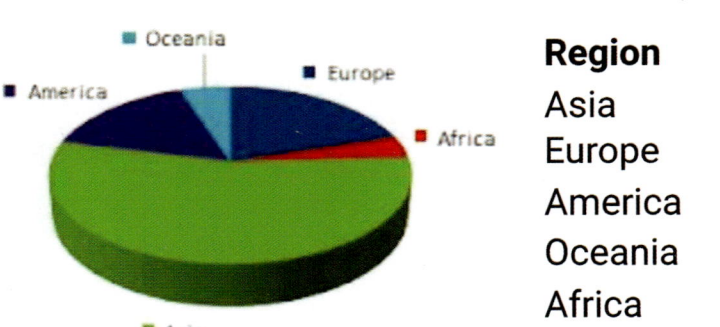

## Aluminum Consumption by Country

| Region | % |
|---|---|
| Asia | 55% |
| Europe | 20% |
| America | 16% |
| Oceania | 5% |
| Africa | 4% |

# Aluminum Uses

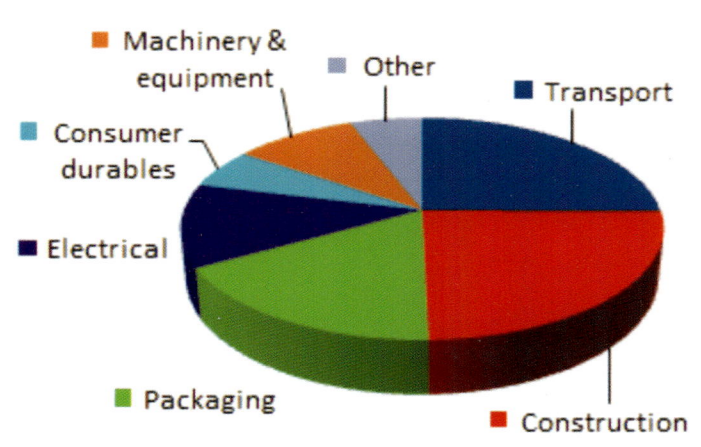

| Market Sector | % |
|---|---|
| Transport | 25% |
| Construction | 25% |
| Packaging | 17% |
| Electrical | 12% |
| Machinery & equipment | 10% |
| Consumer durables | 6% |
| Other | 6% |

## Video Chapter 2 EP 2: Ferrous Alloys & Marine Grade Steels

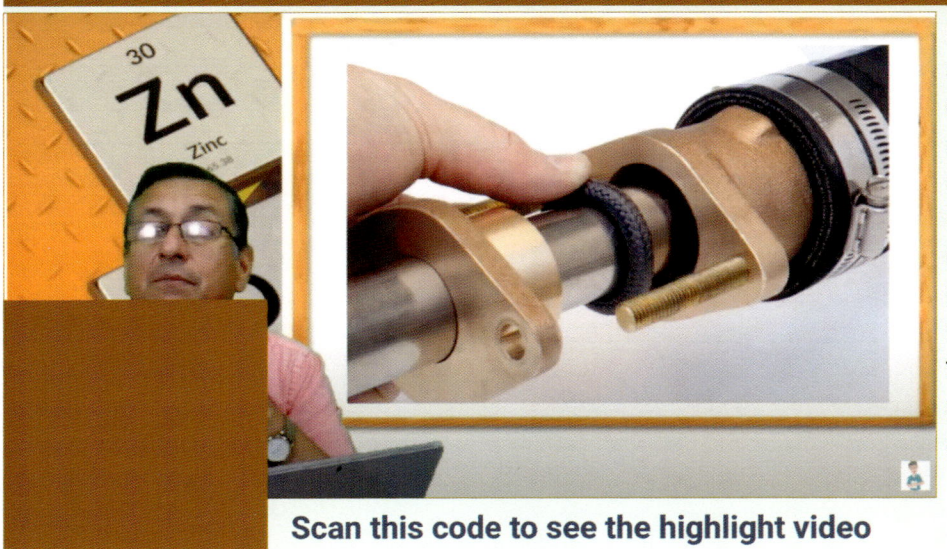

In this video, we are going to study ferrous alloys, how the content of carbon is critical in relation to the quality of the steel, and how other metals such as chromium and nickel change the corrosive and magnetic properties of some marine grade steel alloys. this is a great video for surveyors, marine engineers, and boat owners. Enjoy it!

**Scan this code to see the highlight video**

**Follow me**

Mr. Lopez Classes.com
For Marine Engineers

# Ferrous Alloys

## How They are affected by Corrosion

```
                    ┌──────────────────────┐
                    │  Ferrous Metal Alloys │
                    └──────────────────────┘
```

| Cast Irons | Plain-Carbon Steels | Alloy Steels |
|---|---|---|
| Gray Irons | High-Carbon | Low-Alloy Steels |
| Malleable Iron | Medium-Carbon | HSLA Steels |
| Ductile Iron | Low-Carbon | Microalloyed Steels |
| Compacted Graphite Iron | | Advanced High-Strength Steels |
| Austempered Ductile Iron | | Maraging Steels |
| White Iron | | Stainless Steels |
| | | Tool Steels |

# Carbon Content in Ferrous Alloys

Cast iron contains 2 to 4 percent carbon.
Carbon steels are classified by the percentage of carbon in hundredths of 1 percent they contain.

- Low Carbon Steel: Carbon content up to 0.30 percent
- Medium Carbon Steel: Carbon content from 0.30 to 0.50 percent
- High Carbon Steel: Carbon content from 0.50 to 1.05%

Stainless Steel: The carbon content is held to 0.08% maximum.

# Cast Iron Composition

Cast iron contains high levels of carbon, which makes it a hard, brittle metal. Cast iron was commonly used throughout Europe to make church bells and, in colonial America, pots and pans.

**Cast Iron**

- Carbon 3.5% - 4.0%
- Manganese .5%
- Phosphorous .13%
- Sulfur .13%
- Silicon 1.2%

# Cast Irons

Most cast irons have a chemical composition of 2.5–4.0% **carbon**, 1–3% silicon, and the remainder is iron. It is usually made from pig iron.
Hard skin, softer underneath, but brittle. It corrodes by rusting.

# Grey Cast Iron

Grey cast iron is easily cast but it cannot be forged or worked mechanically either hot or cold.
the carbon content is in the form of flakes distributed throughout the metal.

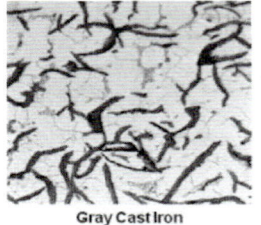
Gray Cast Iron

Grey cast iron is characterized by its graphitic microstructure, which causes fractures of the material to have a grey appearance. It is the most commonly used cast iron.

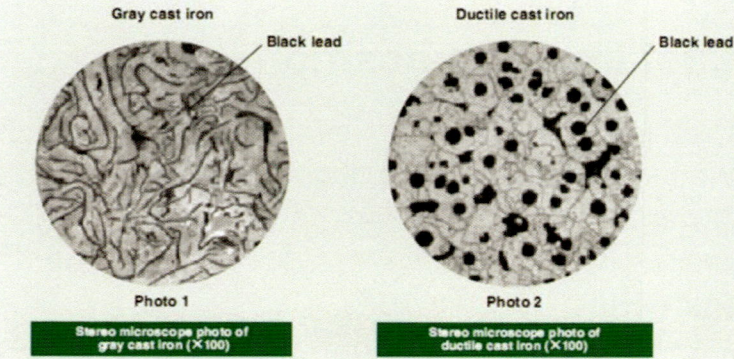

Gray cast iron — Black lead · Ductile cast iron — Black lead

Photo 1 · Photo 2

Stereo microscope photo of gray cast iron (×100) · Stereo microscope photo of ductile cast iron (×100)

# Grey Cast Iron Uses

| Desirable properties | Common Applications |
|---|---|
| Excellent vibration dampening capability | Brake drums, rotors and clutch components |
| High compressive strength | Engine blocks, cylinder liners and heads |
| Good machinability | Pulleys, sheaves and idler wheels |
| Exceptional heat transfer | Gears |
| Low cost alternative to other engineering materials | Hydraulic valves and components |

Marine Technical Training Academy

Mr. Lopez Classes.com
For Marine Engineers

# White Cast Iron

White cast iron has superior tensile strength and malleability. It is also known as **malleable** or **spheroidal graphite** iron.

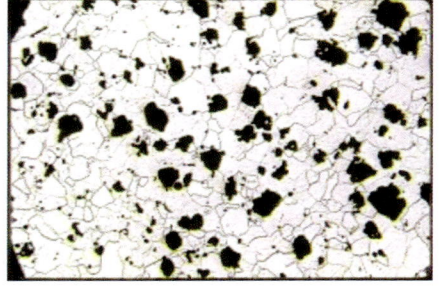

**white cast iron** has carbide impurities which allow cracks to pass straight through.

Since carbide makes up a large fraction of the material, white cast iron could reasonably be classified as a cermet. White iron is too brittle for use in many structural components.

White cast iron has a good hardness and abrasion resistance and is relatively low cost, it finds use in such applications as the wear surfaces Volumes of pumps, Fireplaces, and cooking utensils.

# Carbon Steels

Carbon steels are by far the most produced and used, accounting for about 90 percent of the world's steel production.
High-carbon Steels: with carbon above 0.5 percent.
Medium-carbon steels, with 0.2 to 0.49 percent carbon.
Low-carbon steels, with 0.05 to 0.19 percent carbon.

# Steel Designation

A designation is the specific identification of each grade, type, or class of steel by a number, letter, symbol, name, or a suitable combination.
Chemical composition is by far the most widely used basis for the classification and/or designation of steels.
The most commonly used system of designation in the United States is that of the Society of Automotive Engineers **(SAE)** and the American Iron and Steel Institute **(AISI)**.

# SAE-AISI Designations

The SAE-AISI system is applied to semi-finished forgings, hot-rolled and cold-finished bars, wire rods and seamless tubular goods, structural shapes, plates, sheets, strips, and welded tubing.

The first digit (1), of this designation, indicates carbon steel; i.e., carbon steels comprise 1xxx groups in the SAE-AISI system and are subdivided into four categories due to the variance in certain fundamental properties among them.

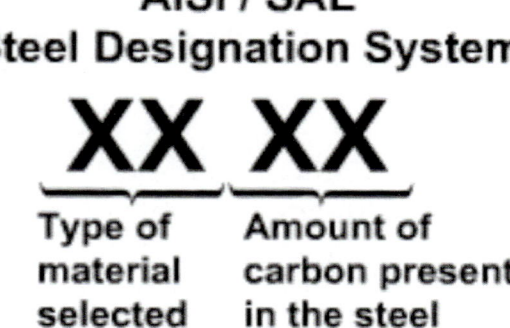

**Video Chapter 2 EP 3: Ferrous Alloys for Marine Applications**

This Video is about the analysis of Ferrous Alloys and how the marine environment affects those alloys. we will study the different types of stainless steel alloys approved for marine environments.
Enjoy it.

Scan this code to see the highlight video

Follow me

# The First Digit

The SAE-AISI system then classifies all other alloy steels using the same four-digit index as follows:

- 2- Nickel steels
- 3- Nickel-chromium steels
- 4- Molybdenum steels
- 5- Chromium steels
- 6- Chromium-vanadium steels
- 7- Tungsten-chromium steels
- 9- Silicon-manganese steels

# SAE-AISI Designations

The plain carbon steels are comprised within the 10xx series (containing 1.00% Mn maximum); resulfurized carbon steels within the 11xx series; resulfurized and rephosphorized carbon steels within the 12xx series; and non-resulfurized high-manganese (up to 1.65%) carbon steels which are produced for applications requiring good machinability are comprised within the 15xx series.

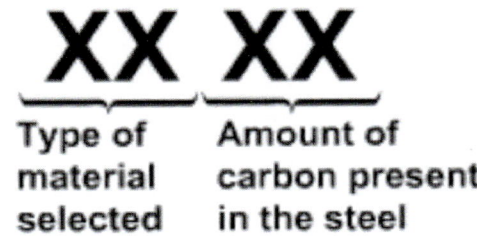

AISI / SAE
Steel Designation System

**XX XX**

Type of material selected | Amount of carbon present in the steel

# SAE-AISI Designations

The second digit of the series indicates the concentration of the major element in percentiles (1 equals 1%). The last two digits of the series indicate the carbon concentration to 0.01%.

SAE 5130 indicates a chromium steel alloy, containing 1% of chromium and 0.30% of carbon.

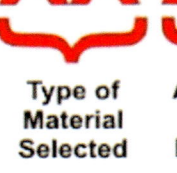

**XX XX Y Z XX**

Type of Material Selected | Amount of Carbon Present in the Steel | Tipo de acero | % Aproximado aleante principal | % Carbono

Mr. Lopez Classes.com
For Marine Engineers

# Higher-Strength

Higher-strength ABS shipbuilding steel comes in six grades of two strengths, **AH32, DH32, EH32, AH36, DH36,** and **EH36.**

The **32** grades have yield strength of 45,500 psi and ultimate tensile strength of 64,000 - 85,000 psi.

The **36** grades have a yield strength of 51,000 psi and ultimate tensile strength of 71,000 - 90,000 psi.

# ABS - Steels

**ABS Steels** are types of <u>structural steel</u> that are standardized by the <u>American Bureau of Shipping</u> for use in <u>shipbuilding.</u>

Ordinary-strength ABS shipbuilding steel comes in a number of grades, **A, B, D, E, DS,** and **CS**. On certified steels.

The <u>yield point</u> for all ordinary-strength ABS steels is specified as between 30,000 and 34,000 <u>psi.</u>

The <u>ultimate tensile strength</u> of ordinary strength alloys is 58,000 - 71,000 psi.

# Mill Scale

**Mill scale,** often shortened to just **scale**, is the flaky surface of <u>hot rolled steel.</u>

Mill scale is composed of iron oxides mostly ferric and is bluish-black in color. It is usually less than 1 mm (0.039 in) thick and initially adheres to the steel surface and protects it from atmospheric corrosion provided no break occurs in this coating.

Mill scale is a porous and poorly adherent scale of magnetite, it will only give limited short-term protection to steel against corrosion.

# Mill Scale for Carbon Steel

Stainless steel differs from carbon steel by the amount of chromium present (Between 10.5 % and 18%).

| Element | Percent by weight |
|---------|-------------------|
| C | 0.20-0.40 |
| Mn | 0.5-1.5 |
| Si | 0.5-1.5 |
| S | 0.05 max |
| P | 0.05 max |
| Cr | 14-18 |
| Ni | 2.0-5.0 |
| Cu | 2.0-4.0 |
| Mo | 1.0 max |
| Nb | 1.5-2.5 |

# Stainless Steel

Stainless steel does not readily corrode, rust or stain with water as ordinary steel does. However, it is not fully stain-proof in low-oxygen, high-salinity, or poor air-circulation environments.

Stainless steels contain sufficient chromium to form a passive film of chromium oxide, which prevents further surface corrosion by blocking oxygen diffusion to the steel surface.

# Chromium (Cr)

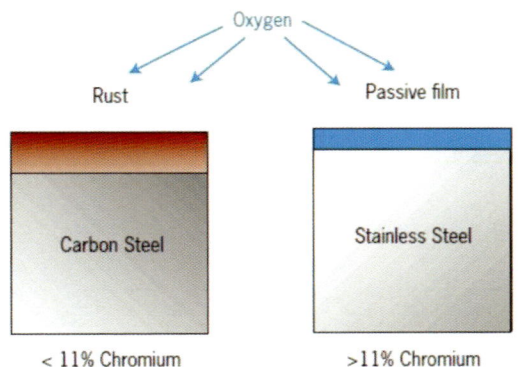

Chromium is added to the steel to increase resistance to oxidation. This resistance increases as more chromium are added.

When added to low alloy steels, chromium can increase the response to heat treatment, thus improving hardenability and strength.

# Manganese (Mn)

Manganese is added to steel to improve hot working properties and increase strength, toughness, and hardenability.

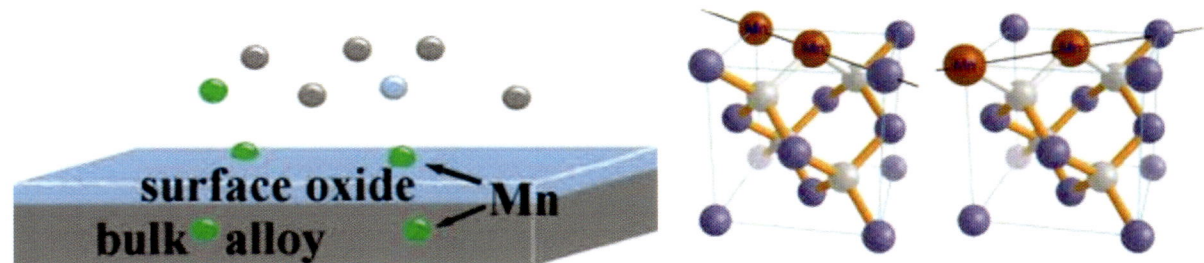

# Nickel (Ni)

Nickel is added in large amounts, over about 8%, to high chromium stainless steel to form the most important class of corrosion and heat resistant steel.

Nickel also improves resistance to oxidation and corrosion. It increases toughness at low temperatures when added in smaller amounts to alloy steels.

STAINLESS     NICKEL
STEEL          PLATED

# Molybdenum (Mo)

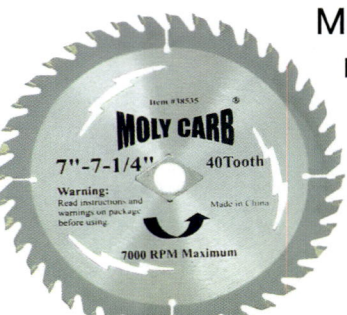

Molybdenum, when added to chromium-nickel steels, improves resistance to pitting corrosion especially by chlorides and sulfur chemicals.

When added to low alloy steels, molybdenum improves high-temperature strengths and hardness.

# Titanium (Ti)

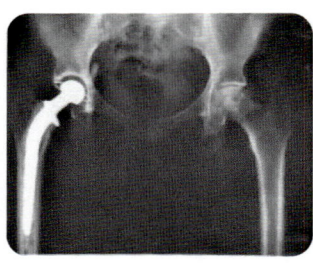

The main use of titanium as an alloying element in steel is for carbide stabilization.

It combines with carbon for titanium carbides, which are quite stable and hard to dissolve in steel, this tends to minimize the occurrence of inter-granular corrosion.

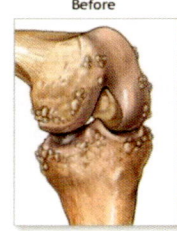
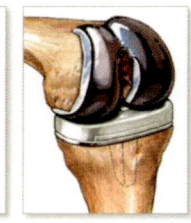

Before     After

# Phosphorus (P)

Phosphorus is usually added with sulfur to improve machinability in low alloy steels, phosphorus, in small amounts, aids strength and corrosion resistance.

Phosphorus additions are known to increase the tendency to crack during welding.

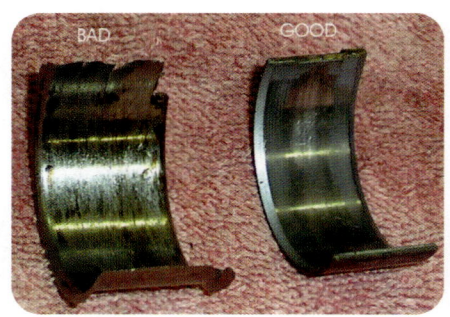

# 18/8 & 18/10 Stainless Steel Grades

The **"grade"** of stainless steel refers to its quality, durability, and temperature resistance. The **numbers** (18/8, 18/10, etc.) are the composition of the stainless steel and refer to the amount of chromium and nickel (respectively) in the product.
**18/8 and 18/10**: These are the two most common grades of stainless steel used for food preparation and dining, also known as Type 304 (**304 Grade**), and are part of the 300 series.

The first number,18, refers to the amount of chromium present and the second represents the amount of nickel.
**18/0** - Contains a negligible amount of nickel (0.75%) and therefore has reduced corrosion resistance.
18/0 is also referred to as **Type 430**, is part of the **400 series** and, unlike 300 series stainless steel, **is magnetic.**

# 304 Series Stainless

Type 304 is used for sinks, tabletops, coffee urns, stoves, refrigerators, milk and cream dispensers, and steam tables. It is also used in numerous other utensils such as cooking appliances, pots, pans, and flatware.
It is immune to foodstuffs, sterilizing solutions, most organic chemicals and dyestuffs, and a wide variety of inorganic chemicals.

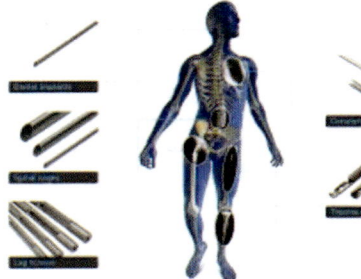

# 316 Series Stainless

**Marine-grade stainless**, or SAE 316 stainless steel, is molybdenum-alloyed steel.
It is the preferred steel for use in marine environments

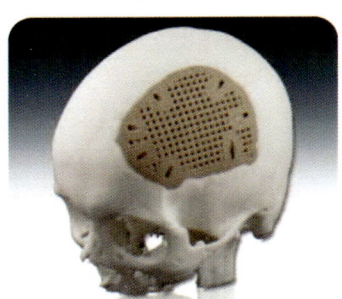

because of its greater resistance to pitting corrosion than other grades of steel.
Type 316 is also used extensively for surgical implants within the hostile environment of the body.

## Stabilized Grades

During the welding of Stainless, the carbon in the steel reacts with the chromium, leaving the welding area exposed to intergranular corrosion.

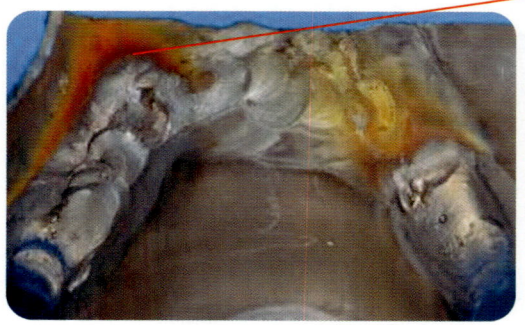

The stainless steel grades L (Low level of Carbon) were developed to minimize this inconvenience ( 304L,316L, and 317L). These grades have lower levels of carbon.

## Other Common series

### TYPE 316

- stainless steel containing 2%-3% molybdenum (whereas 304 has none). The inclusion of molybdenum gives 316 greater resistance to various forms of deterioration.

### TYPE 409

- stainless steel suitable for high temperatures. This grade has the lowest chromium content of all stainless steel and thus is the least expensive.

### TYPE 410

- It is a low-cost, heat-treatable grade suitable for non-severe corrosion applications.

## TOPICS

### Video Chapter 3 EP 1: Oxidation and Corrosion in Marine Environments

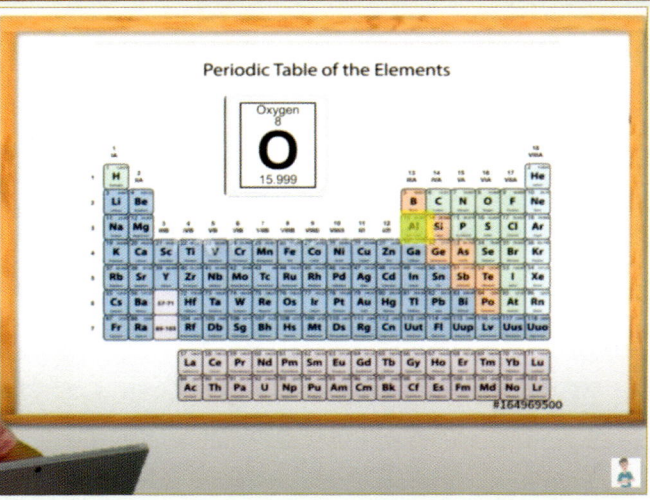

In this Video, we are going to analyze what is the meaning of oxidation and how cumulative oxidation produces corrosion. This video is a great tool for marine engineers, marine surveyors, and inspectors. Enjoy it.

Scan this code to see the highlight video

Follow me

# Oxidation States of Metals

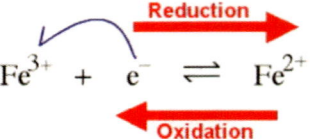

The oxidation state of an element is related to the number of electrons that an atom loses, gains, or appears to use when joining with another atom in compounds.

It also determines the ability of an atom to oxidize (to lose electrons) or to reduce (to gain electrons).

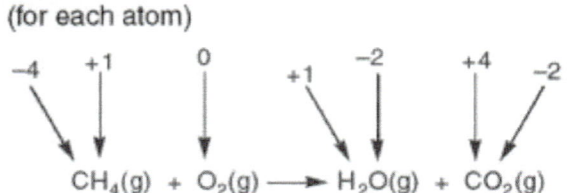

# Oxidation

Is the interaction of oxygen molecules with other metals that are less electronegativity.

'Oxidation' is the process of either stealing or donating electrons, when the electrons are stolen or donated, the metal atoms will dissolve in the water.

# The Process

The majority of pure metals come from oxides, salts, or hydroxides. For that reason, they just need some atoms of hydrogen and oxygen to return to their original form.

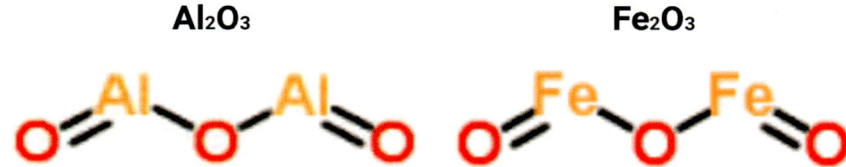

# Oxidants and Reducers

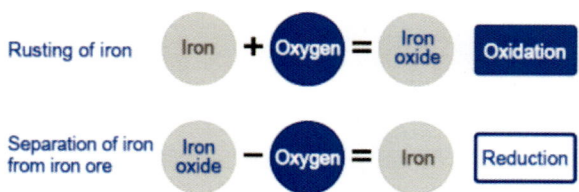

An **Oxidant agent**, or **oxidant**, gain electrons and is reduced in a chemical reaction

A **reducing agent**, or **reductant**, loses electrons and is oxidized in a chemical reaction.

# Oxidants and Reducers

For example, oxygen (O) and fluorine (F) are very strong oxidants. On the other hand, lithium (Li) and sodium (Na) are incredibly strong reducing agents (likes to be oxidized), meaning that they easily lose electrons.

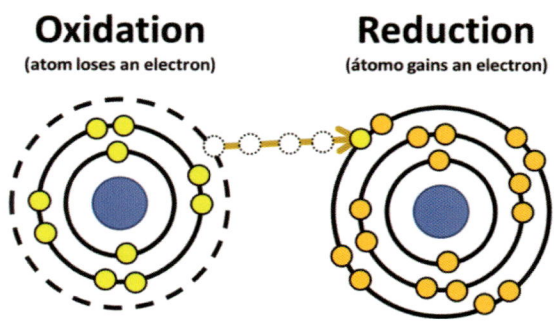

**Oxidation**
(atom loses an electron)

**Reduction**
(átomo gains an electron)

## Reduction & Oxidation

*Redox Rxn: Electron transfer from X to Y*

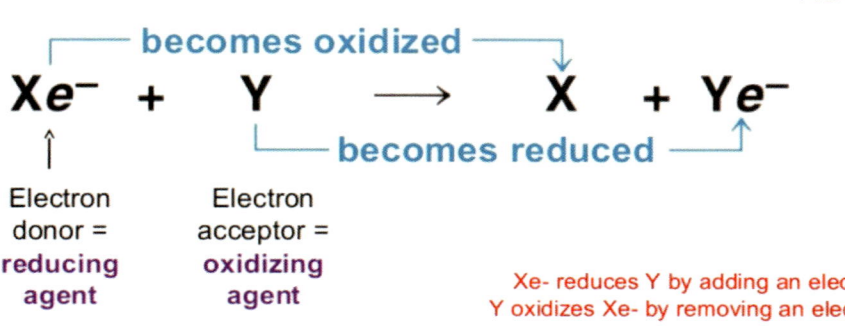

becomes oxidized

$$Xe^- + Y \longrightarrow X + Ye^-$$

becomes reduced

Electron donor = **reducing agent**

Electron acceptor = **oxidizing agent**

Xe- reduces Y by adding an electron to it.
Y oxidizes Xe- by removing an electron from it.

The electron donor is called "The Reducing Agent"

## Reactants & Products

Reactants are substances that start a chemical reaction. Products are substances that are produced in the reaction.

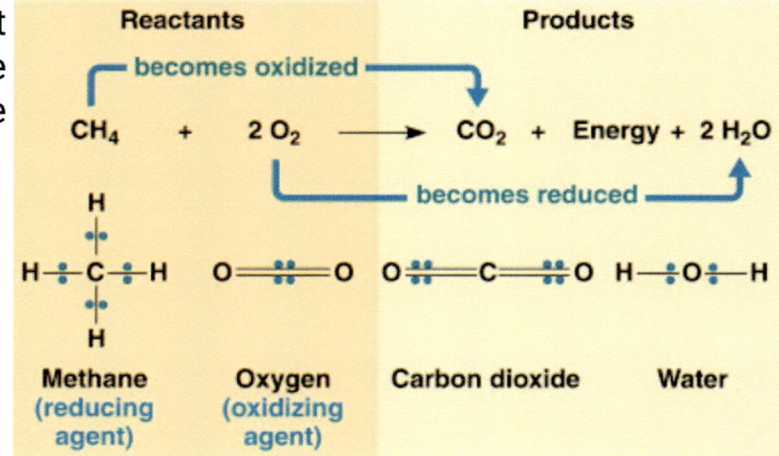

Reactants → Products

becomes oxidized

$$CH_4 + 2 O_2 \longrightarrow CO_2 + Energy + 2 H_2O$$

becomes reduced

Methane (reducing agent)    Oxygen (oxidizing agent)    Carbon dioxide    Water

# Biodiesel Formula

## Oxidation or Corrosion?

Corrosion is the chemical or electrochemical reaction that causes an engineered material to disintegrate as a reaction to its surroundings.
It is a gradual process with the elements eating away at the materials making them deteriorate and break up because of the oxidation of the metals as a chemical reaction to an oxidant, usually oxygen.
Oxidation is the process where electrons (which bind atoms together to create materials) are drawn away by free oxygen molecules which are relatively unstable and looking for available electrons.

## What are Oxides ?

An oxide is a chemical compound of oxygen with other chemical elements.
Oxides containing only one oxygen are called oxide or monoxide, those containing two oxygen atoms of dioxide.

# Types of Oxides

There are two types of oxides: Basic Oxides and Acidic Oxides.

Acidic Oxide = Oxygen + Non-Metal ($NO_2$, $SO_3$)

1. Do not react with acids.
2. React with bases and alkalis to form salt & water.
3. Dissolve in water to form acidic solutions. 4. Usually gases at room temp.

Basic Oxide = Oxygen + Metal ($MgO$, $FeO$)

1. Do not react with bases.
2. React with acids to form salt & water.
3. Basic Oxides are usually insoluble in water. Those that dissolve in water form alkaline solutions.

# Acidic Oxides

An oxide that combines with water to give an acid is termed as an **acidic oxide.**
Acidic oxides are the oxides of **non-metals** (groups 4A to 7A). These acidic oxides form acids with water:

- Sulfurous Acid
  - $SO_2 + H_2O \longrightarrow H_2SO_3$
- Carbonic Acid
  - $CO_2 + H_2O \longrightarrow H_2CO_3$
- Sulfuric Acid
  - $SO_3 + H_2O \longrightarrow H_2SO_4$

Acidic oxides are inorganic chemicals that react with <u>water</u> to form an <u>acid.</u>
Acidic oxides are generally formed by non-metals.
For example, carbon dioxide reacts with water to produce carbonic acid:

$$H_2O + CO_2 \longleftrightarrow H_2CO_3$$

# Basic Oxides

The oxide that gives a base in water is known as a basic oxide.
Generally, <u>Group 1</u> and <u>Group 2</u> elements form bases called base anhydrides or basic oxides e.g.,
$K_2O(s) + H_2O(l) \longrightarrow 2KOH(aq)$.
Basic oxides are the oxides of <u>metals</u>. If soluble in water they react with water to produce hydroxides (alkalies) e.g.,

- $CaO + H_2O \longrightarrow Ca(OH)_2$
- $MgO + H_2O \longrightarrow Mg(OH)_2$
- $Na_2O + H_2O \longrightarrow 2NaOH$

A **basic oxide** is an oxide that shows basic properties in opposition to <u>acidic oxides</u> and that either.
- reacts with <u>water</u> to form a <u>base</u>, or
- reacts with an <u>acid</u> to form a <u>salt</u> and water.

Examples include:
- <u>Sodium oxide</u>, which reacts with water to produce <u>sodium hydroxide</u>
- <u>Magnesium oxide</u>, which reacts with <u>hydrochloric acid</u> to form <u>magnesium chloride</u>
- <u>Copper(II) oxide</u>, which reacts with <u>nitric acid</u> to form <u>copper nitrate</u>

Basic oxides are oxides mostly of <u>metals,</u> especially <u>alkali</u> and <u>alkaline earth</u> metals.

A general property of metal oxides is that they tend to react with water to form basic solutions of the metal hydroxide
For example, MgO is a basic oxide. It dissolves in water to form magnesium hydroxide, $Mg(OH)_2$. The solution contains "1" $Mg^{2+}$ ion for every "2" hydroxide ions, $OH^{1-}$
This means one formula unit forms the following ions. $Mg^{2+}$ $OH^{1-}$ $OH^{1-}$ where the total electric charge in the solution adds to zero.

## Examples of basic oxides

| Basic Oxide | Formula |
|---|---|
| magnesium oxide | $MgO$ |
| sodium oxide | $Na_2O$ |
| calcium oxide | $CaO$ |
| copper(II) oxide | $CuO$ |

# Acids

Acids are compounds that contain Hydrogen (Hydrochloric, HCl; Sulphuric, $H_2SO_4$; Nitric, $HNO_3$)
However, not all compounds that contain Hydrogen are acids (Water, $H_2O$; Methane, $CH_4$).

| | |
|---|---|
| HCl<br>Hydrochloric acid | $HC_2H_3O_2$<br>Acetic acid |
| $H_2SO_4$<br>Sulfuric acid | $H_3C_6H_5O_7$<br>Citric acid |
| $HNO_3$<br>Nitric acid | $H_3PO_4$<br>Phosphoric acid |
| $H_2CO_3$<br>Phosphoric acid | $H_2C_2O_4$<br>Oxalic acid |

# Organic and Inorganic Acids

**Organic vs. Inorganic**

- Organic compounds mean that **carbon** is contained.
- Inorganic compounds mean that **no carbon** is contained

*There are a few exceptions...*

An organic acid is an organic substance that has the properties of an acid.

inorganic acids, also known as mineral acids, come from inorganic substances. Some examples of inorganic acids include sulphuric acid, hydrochloric acid, nitric acid, boric acid, and hydrofluoric acid.

Organic acids occur in or can be produced from, animal and vegetable matter. In addition to hydrogen, organic acids always contain carbon and at least one other element.
Inorganic acids, which contain no carbon, are sometimes called mineral acids. The most important inorganic acids are sulfuric, phosphoric, and nitric.

**Inorganic vs Organic**

| | |
|---|---|
| ·Non-living<br>·Do not contain carbon (exceptions)<br>·Needed for life<br>·Plants organize inorganic compounds into organic<br>·$H_2O$, $O_2$, $CO_2$<br>·Minerals, nitrates | ·Produced by living things<br>·All contain carbon<br>·Important organic compounds<br>   –Carbohydrates<br>   –Lipids<br>   –Proteins<br>   –Nucleic acids |

## Video Chapter 3 EP 2: Ceramics - Metal Oxides and Fiberglass Oxidation

In this video, we are going to analyze the ceramics and metals oxides and their applications in boat hull coatings and antifouling protection.

Scan this code to see the highlight video

**Follow me**

# Ceramics

Ceramics are generally thought of as inorganic and nonmetallic solids with a range of useful properties, including very high hardness and strength, extremely high melting points, and good electrical and thermal insulation.

Table I Comparison between the properties of ceramics and metals

| Comparison property | Ceramics | Metals |
| --- | --- | --- |
| Corrosion resistance | Good | Fair |
| Creep resistance | High | Fair |
| Formation | Fair (sintering) | Easy (forging) |
| Fracture toughness | Difficult | Easy |
| Joining | Difficult | Easy |
| Oxidation resistance | Good | Fair |

Source: Lovatt et al. (2009)

# What are Ceramics ?

Advanced ceramics are not generally clay-based. Instead, they are either based on oxides or non-oxides or combinations of the two.
- Typical oxides used are alumina ($Al_2O_3$) and zirconia ($ZrO_2$).
- Non-oxides are often carbides, borides, nitrides, and silicides, for example, boron carbide ($B_4C$), silicon carbide ($SiC$), and molybdenum disilicide ($MoSi_2$).

The best-known ceramics are pottery, glass, brick, porcelain, and cement. But the general definition of a ceramic—a nonmetallic and inorganic solid.

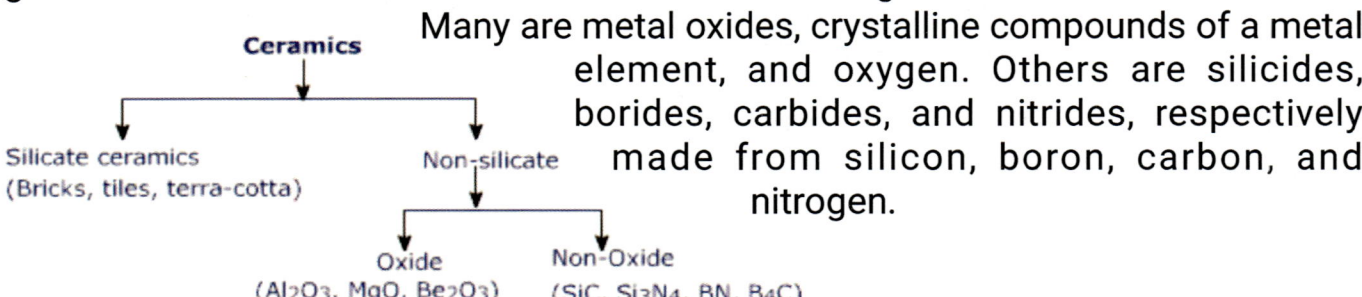

Many are metal oxides, crystalline compounds of a metal element, and oxygen. Others are silicides, borides, carbides, and nitrides, respectively made from silicon, boron, carbon, and nitrogen.

# Ceramics Properties

In general, ceramics are corrosion-resistant and hard, but brittle. Most ceramics are also good insulators and can withstand high temperatures.
Traditional ceramics are used in dishes, crockery, flowerpots, and roof and wall tiles.

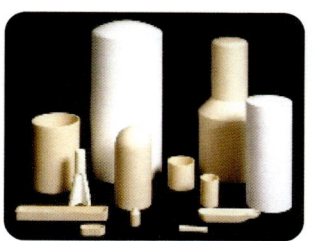

# Aluminum Oxide

Everyone knows that if you leave Aluminum out in the rain, it rusts. If it rusts for a long period of time, the metal disappears and you're left with a pile of white powder-rust or Aluminum oxide, which has the same composition as **Alumina** or $Al_2O_3$.

# $Al_2O_3$ Ceramic Properties

Alumina is one of the most cost-effective and widely used materials in the family of engineering ceramics.
Typical Uses
*   High-Temperature Electrical Insulator
*   High Voltage/Current Insulator
*   Seal Rings
*   Wear Pads

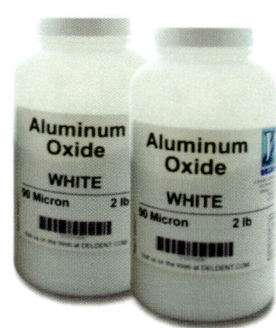

# Copper Oxide Properties

Oxygen can combine with copper can combine in different ways to form two types of oxides:
copper(I) oxide ($Cu_2O$), which is normally a reddish powder.
*   it is used in photoelectric cells and light detectors
*   A ceramic material made from it acts as a superconductor at relatively high temperatures
*   It is an ingredient in many fungicides

copper(II) oxide, also known as cupric oxide ($CuO$) which is usually a black powder.
*   $CuO$ is sometimes added to clay glazes as a pigment
*   Is used as an abrasive for polishing lenses and other optical components.

## Copper Oxide Colors

Copper turns green when exposed to the atmosphere due to the reaction with Oxygen.

The major factors that control the initial rate of attack on copper, and that cause copper to turn green, are moisture, temperature, and the level of pollution.

Soon after exposure of copper to the atmosphere, due to the fact that copper oxidize, the bright copper surface takes on a dull tan tarnish.

## Copper Oxide Applications

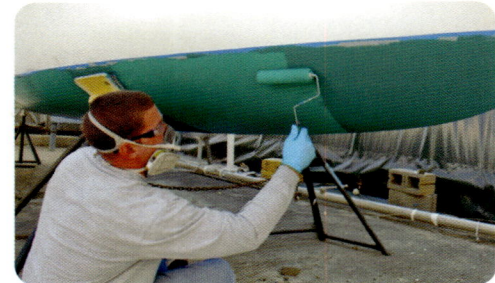

Copper oxide is the most popular biocide used in antifouling paints today.

Antifouling uses of copper oxide include commercial and non-commercial applications for boat and ship hulls and miscellaneous applications such as underwater structures and piers.

## The Iron Ore

As the Iron ore is mined out of the ground, its chemical composition is two iron atoms bonded with three oxygen atoms.

## Iron Oxidation

If you leave iron out in the rain, it rusts. After some days exposed to the environment the metal

$$Fe_2O_3$$

Iron Oxide Powder

disappears and you're left with a pile of brownish-red powder-rust or iron oxide, which has the same composition as iron ore.

Marine Technical Training Academy

Mr. Lopez Classes.com
For Marine Engineers

# Iron Oxides

Iron Oxide is a mineral, colored black to steel or silver-gray, brown to reddish-brown, or red. It is mined as the main ore of Iron.
Iron oxides are used as core sand additives to improve sand quality.
Iron oxides are also used in exothermic, welding rods, offshore drilling muds, and colorants in both building products and brown glass beverage bottles.

There are sixteen known iron oxides.
Iron oxides and oxide-hydroxides are widespread in nature and are widely used by humans, e.g., as iron ores, pigments, and hemoglobin.

$$Fe_2O_3 \quad FeO \quad Fe_3O_4$$

iron (III) oxide    iron (II) oxide

Iron oxides are widely used as inexpensive, durable pigments in paints, coatings and colored concretes.

# Fiberglass Oxidation

This process of oxidation, which is similar to rust on metal, takes the shine off the gel coat top layer on fiberglass products, leaving a gritty residue.

Oxidation is the chalky, porous appearance that is the result of unprotected and/or neglected Gelcoat or paint. This process is mainly caused by damage from UV rays which slowly breaks down unprotected gelcoat and other surfaces.
If you have worked with gel coats and paints before, you know that unlike automotive clearcoats, oxidation happens much faster because it is much more porous, and keeping it protected is critical to maximizing its life span and appearance. It requires select cleaning, protective products, and equipment to bring back and maintain its luster.

Waxing and polishing only coat the oxidation, allowing it to continue under the wax layer. The solution involves scrubbing the fiberglass with a specially formulated oxidation remover, available at outdoor suppliers and boat centers, then coating the surface with a sealant.

Mr. Lopez
Classes.com
For Marine Engineers

## How to Prevent Oxidation in Fiberglass

The most practical way is regular cleaning, polishing, and waxing. Cleaning is the most logical and cost-effective of all, especially when you want to preserve your Gelcoat and keep oxidation at bay.

One of the best things to do is wash and dry your boat each time you pull it out of the water or use it. That will prevent the dirty water from staining your fiberglass colors, corrosive salt buildup, and mold from settling into the surfaces.

Also, I recommend using appropriate hardware (only marine grade alloys) for hinges, brackets, handrails, screws and supports to avoid the

progress of corrosion (Crevice corrosion) around those metallic elements.

## Removing Fiberglass Oxidation

Saturate a sponge brush with liquid oxidation remover. Work outdoors to minimize mess and avoid breathing chemical fumes.

Scrub the fiberglass surface with the sponge brush, adding more oxidation remover as necessary until the fiberglass gel coat begins to resemble its original appearance. Let the surface dry completely.

Apply the fiberglass sealer directly to the cleaned surfaces using the sponge applicator on the product bottle. Wait for the surface to dry completely.

Buff the treated fiberglass with an electric buffer until the surfaces shine. You can now apply wax, if desired, to add an additional layer of protection now that the oxidation is stripped away and the surface is resealed.

## The Marine Environment

## TOPICS

## Video Chapter 4 EP 1: The Marine Environment

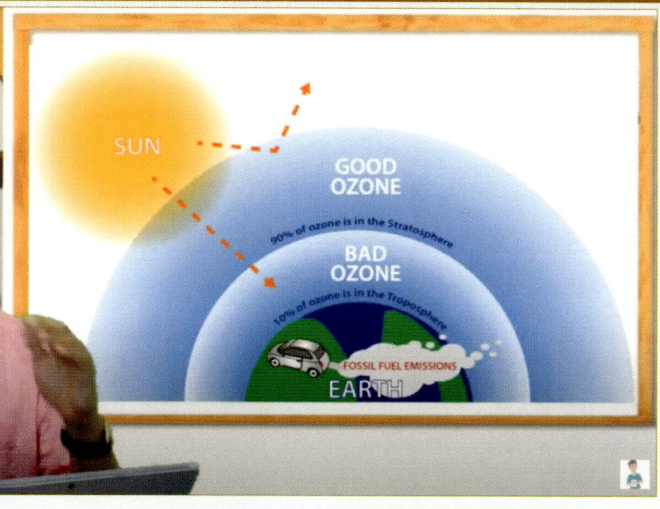

In this Video, we are going to analyze how the marine affects the process of corrosion. How the pollutants produced in the incomplete combustion of the Gas and Diesel engines accelerate the corrosion process. Enjoy it

**Scan this code to see the highlight video**

Follow me

# Air Composition

Air is a mixture of gases - 78% nitrogen and 21% oxygen - with traces of water vapor, carbon dioxide, argon, and various other components.

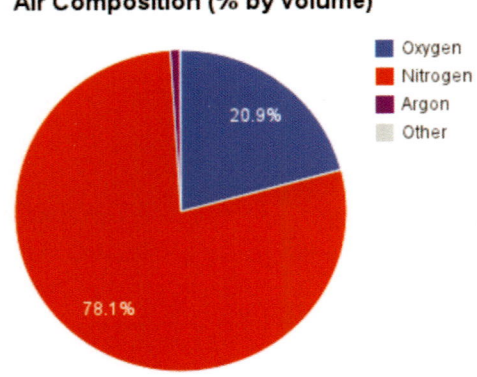

Air Composition (% by volume)

# Other Components in Air

Sulfur dioxide - $SO_2$ - 1.0 parts/million (ppm)
Methane - $CH_4$ - 2.0 parts/million (ppm)
Nitrous oxide - $N_2O$ - 0.5 parts/million (ppm)
Ozone - $O_3$ - 0 to 0.07 parts/million (ppm)
Nitrogen dioxide - $NO_2$ - 0.02 parts/million (ppm)
Iodine - $I_2$ - 0.01 parts/million (ppm)
Carbon monoxide - $CO$ - 0 to trace (ppm)
Ammonia - $NH_3$ - 0 to trace (ppm)

# Unpolluted Air

The atmosphere surrounding the Earth is a mixture of gases. In some places, human activities have added other gases to the atmosphere, which are called pollutants.
The atmosphere also contains a small but important amount of carbon dioxide, approximately 0.04 percent, and tiny amounts of a few other gases.
The burning of fuels releases a large amount of carbon dioxide into the atmosphere. This is thought to cause global warming.

# Pollutant Gases

Other pollutant gases which are released into the air when fuels are burned include carbon monoxide, nitrogen oxides, and sulfur dioxide.
The incomplete combustion of fuels also releases small particles of solids, such as carbon, into the air. This makes buildings dirty and affects the lungs.

**Emissions of Primary Air Pollutants**

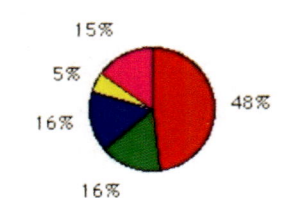

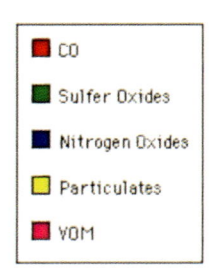

# Air Pressure

Air pressure decreases with altitude. At sea level, air pressure is about 14.7 pounds per square inch (1 kilogram per square centimeter). At 10,000 feet (3 km), the air pressure is 10 pounds per square inch (0.7 kg per square cm). There is also less oxygen to breathe.

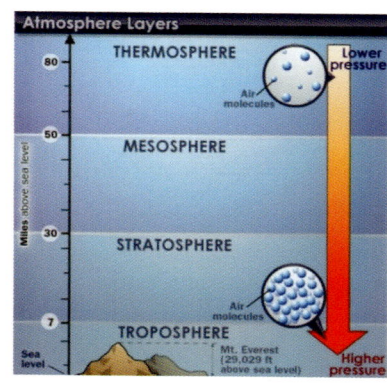

# Atmosphere Layers

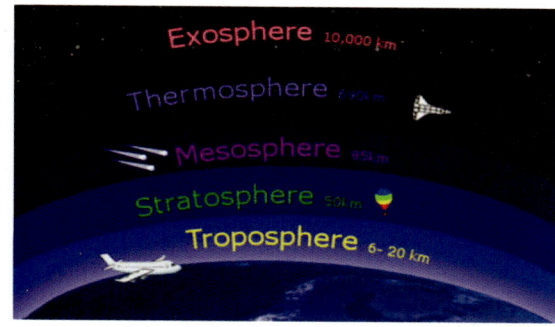

Earth's atmosphere is divided into five main layers, the Exosphere, the Thermosphere, the Mesosphere, the Stratosphere, and the Troposphere.

## The Troposphere

The **Troposphere** is the layer closest to Earth's surface. It is 4 to 12 miles (7 to 20 km) thick and contains half of Earth's atmosphere. Air is warmer near the ground and gets colder higher up.

Nearly all of the water vapor and dust in the atmosphere are in this layer and that is why clouds are found here.

# The Stratosphere

The **stratosphere** is the second layer. It starts above the troposphere and ends about 31 miles (50 km) above ground. Ozone is abundant here and it heats the atmosphere while also absorbing harmful radiation from the sun.

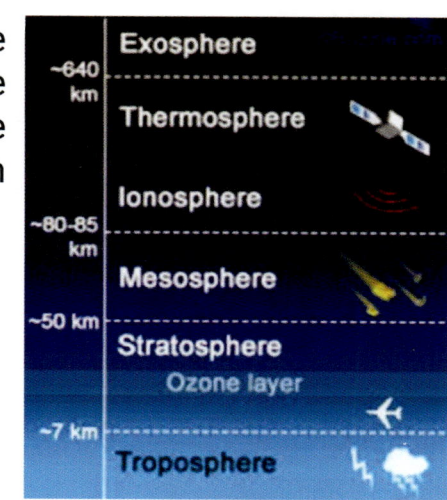

## What is Ozone?

Ozone (O3) is a highly reactive gas composed of three oxygen atoms.

It is both a natural (Good Ozone) and a man-made (Bad Ozone) product that occurs in the Earth's upper atmosphere (the stratosphere) and lower atmosphere (the troposphere).

Ozone molecules are much less stable than regular (O2) oxygen molecules. They are very prone to react chemically with other substances.

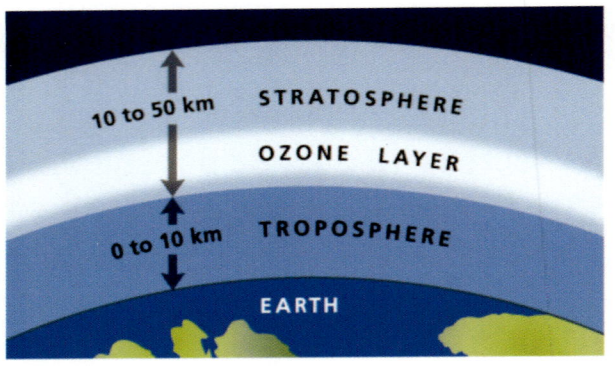

## How The Ozone React

Ozone is ready to react with whatever it meets. This makes it very useful for cleaning and disinfecting. But, when it comes in contact with living tissues like our lungs it can cause damage and illness.

 Ozone can also corrode building materials, statues and monuments, and natural rock features in the landscape.

## The Ozone

Nobody puts ozone directly into the air. Instead, cars and trucks, gas stations, and factories put the ingredients for ozone into the air every day.

Depending on where it is in the atmosphere, ozone affects life on Earth in either good or bad ways.

# Stratospheric Ozone (Good Ozone)

Stratospheric ozone is formed naturally through the interaction of solar ultraviolet (UV) radiation with molecular oxygen (O2). The "ozone layer," approximately 6 through 30 miles above the Earth's surface, reduces the amount of harmful UV radiation reaching the Earth's surface.

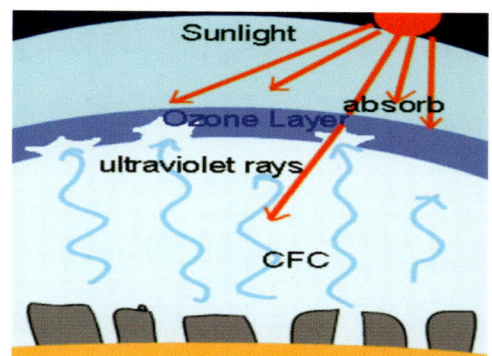

# Tropospheric Ozone (Bad Ozone)

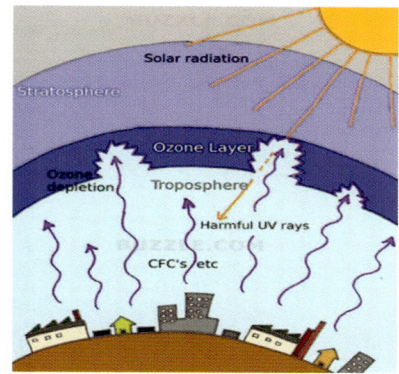

Tropospheric or ground-level ozone – what we breathe – is formed primarily from photochemical reactions between two major classes of air pollutants, volatile organic compounds (VOC) and nitrogen oxides (Nox).

## Ozone-Depleting Substances ODS

Other ozone-depleting substances (ODS), include hydrochlorofluorocarbons (HCFCs), and volatile organic compounds (VOCs). These are often found in vehicle emissions, byproducts of industrial processes, refrigerants, and aerosols.
ODS is relatively stable in the lower atmosphere of the Earth, but in the stratosphere, they are exposed to ultraviolet radiation, and thus, they break down to release a free chlorine atom.

# Causes of Ozone Depletion

This free chlorine atom reacts with an ozone molecule (O3) and forms chlorine monoxide (ClO), and a molecule of oxygen. Now, ClO reacts with an ozone molecule to form a chlorine atom and two molecules of oxygen. The free chlorine molecule again reacts with ozone to form chlorine monoxide. The process continues, and this results in the depletion of the ozone layer.

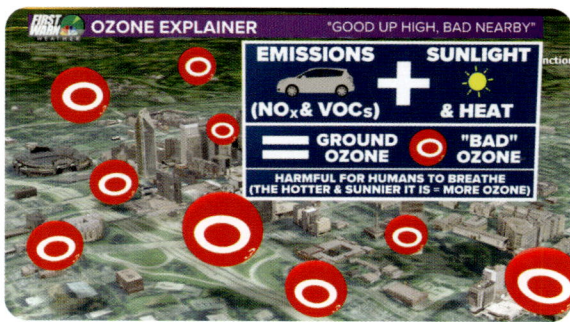

## Marine Corrosion

# TOPICS

**Video Chapter 5 EP 1: Marine Corrosion Alkalinity, Acidity and Ph Level**

In this Video we will analyze how external factors such as : sea water alkalinity , acidity , salinity and the Ph level affect the corrosion process in your boat.

**Scan this code to see the highlight video**

**Follow me**

# What is Corrosion

Is the chemical deterioration of a metal, an alloy of metals, or a compound material. Corrosion is the disintegration of metal through an unintentional chemical or electrochemical action, starting at its surface.

In other words, Corrosion is a continuous and advanced oxidation process

All metals exhibit a tendency to be oxidized, some more easily than others.

Knowledge of a metal's location in the periodic table and the selection of appropriate marine grade alloys are important pieces of information to design and install marine structures.

# What is Marine Corrosion?

Saltwater is a much better conductor than fresh, so the rate of corrosion increases dramatically.

Total removal of Chlorine and salt is the only way to stop this process as even the smallest amount of residual salt will keep the reaction going.

Metals such as stainless steel, titanium, manganese bronze, and anodized aluminum minimize this destruction by forming a thin, protective oxide layer covering their surface. This oxide surface is self-healing when scratched, thereby minimizing the surface degradation.

# Factors Affecting Corrosion

The most important factor in atmospheric corrosion, overriding pollution or lack of it, is moisture, either in the form of rain, dew, condensation, or high relative humidity (RH). In the absence of moisture, most contaminants would have little or no corrosive effect.

The corrosion process increase under the presence of currents. Generally, those currents are small DC currents coming from "static electricity" stored in appliances, engine blocks and, metallic devices. Also from sudden "shorts to ground.

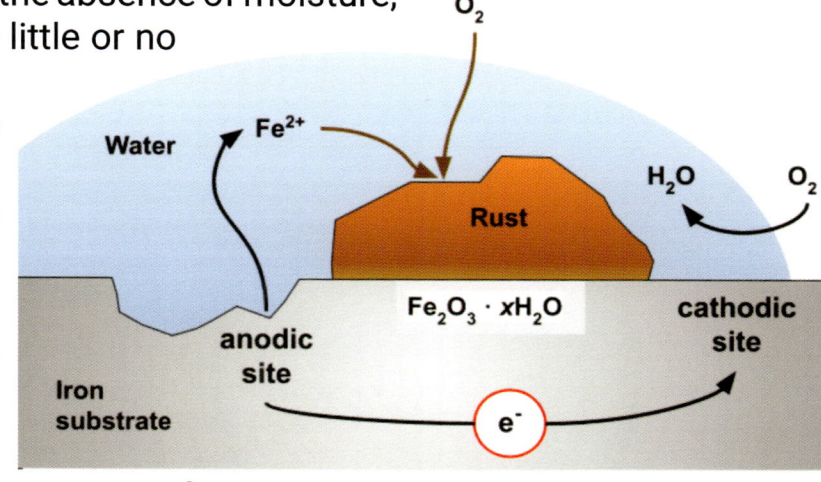

$Fe(s) \rightarrow Fe^{2+}(aq) + 2e^-$          $O_2(g) + 4H^+(aq) + 4e^- \rightarrow 2H_2O(l)$

# Definitions

**Dew** is water in the form of droplets that appears on thin, exposed objects in the morning or evening due to condensation.

**Water stagnation** occurs when water stops flowing.

An **alkaline solution** is a mixture of base solids dissolved in water. The potential of hydrogen, also known as the pH scale, measures the alkalinity or acidity level of a solution.

# Relative Humidity (RH)

RH is defined as the ratio of the quantity of water vapor present in the atmosphere to the saturation quantity at a given temperature, and it is expressed as %.

The critical humidity level is a variable that depends on the nature of the corroding material, the tendency of corrosion products and surface deposits to absorb moisture, and the presence of atmospheric pollutants.

# Factors Affect Corrosion

Marine corrosion is influenced by many water characteristics, by the metals used, and by any stray electrical current.

Regarding the water, primary factors include alkalinity, hardness, and pH, which can also influence corrosion.

Alkalinity, hardness, and pH interact to determine whether the water will produce scale or corrosion or will be stable.

A film of dew, saturated with sea salt or acid sulfates, and acid chlorides of a marine atmosphere provides an aggressive electrolyte for the promotion of corrosion.

# Corrosion Vs Humidity

In the humid Tropics where nightly condensation appears on many surfaces, the stagnant moisture film either becomes alkaline from reaction with metal surfaces or picks up carbon dioxide and becomes aggressive as a dilute acid.

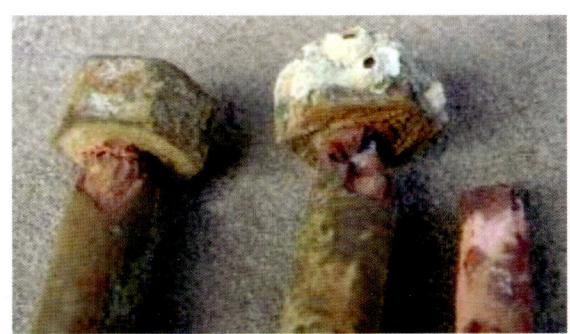

# Aqueous Corrosion

The term 'aqueous corrosion' describes the majority of the most troublesome problems encountered in contact with seawater.

Corrosion by seawater, aqueous corrosion, is an electrochemical process, and all metals and alloys when in contact with seawater have a specific electrical potential (or corrosion potential) at a specific level of seawater acidity or alkalinity - the pH.

# How the pH level affect Corrosion

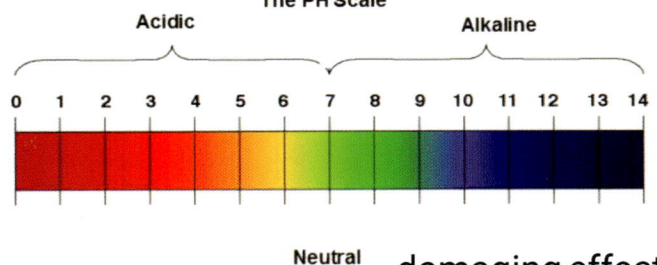

Both acids and alkalies have the capability of being corrosive, although one would have a pH range of 0 (acid), while the other would range in the area of 14 (alkali).

Sodium hydroxide, a very strong and corrosive alkali would have the same damaging effect on human tissue as sulfuric acid.

# Methods to Determine the pH

One of these is using a piece of pH indicator paper. When the paper is pushed into a solution it will change color. Each different color indicates a different pH value.

The pH value is used to represent the acidity of a solution.

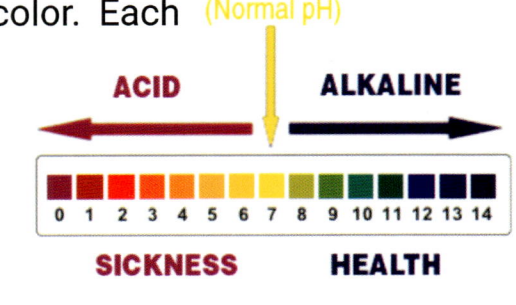

## pH Level

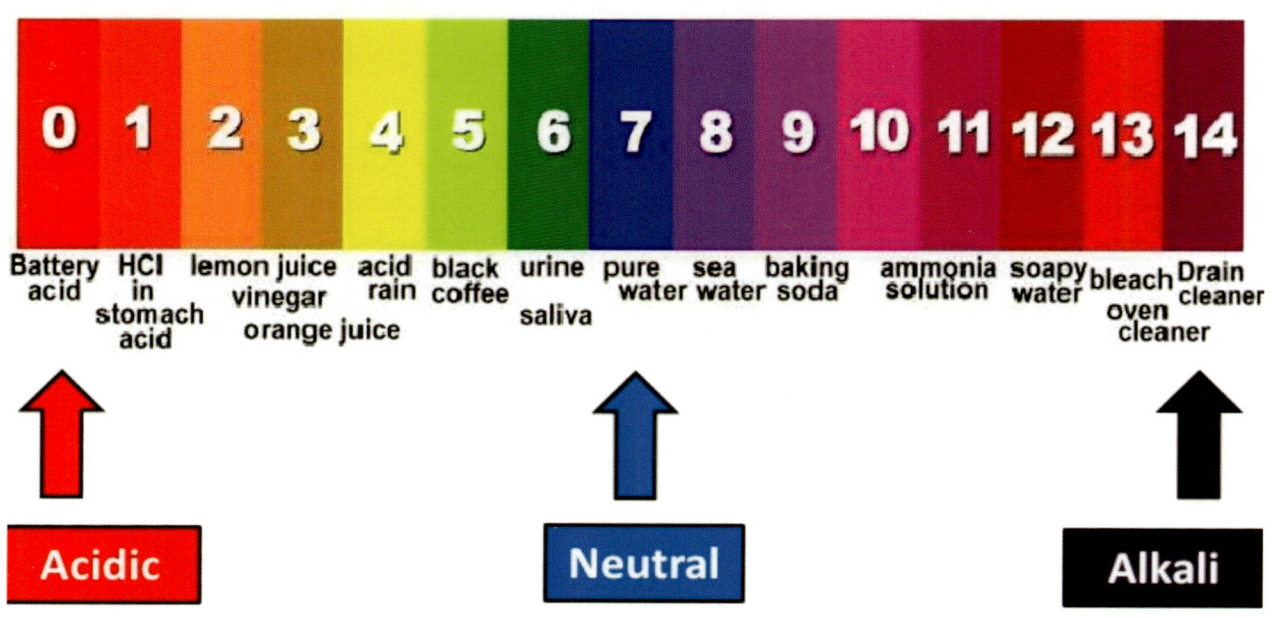

# PH SCALE

The pH scale is used to express acidity and alkalinity.

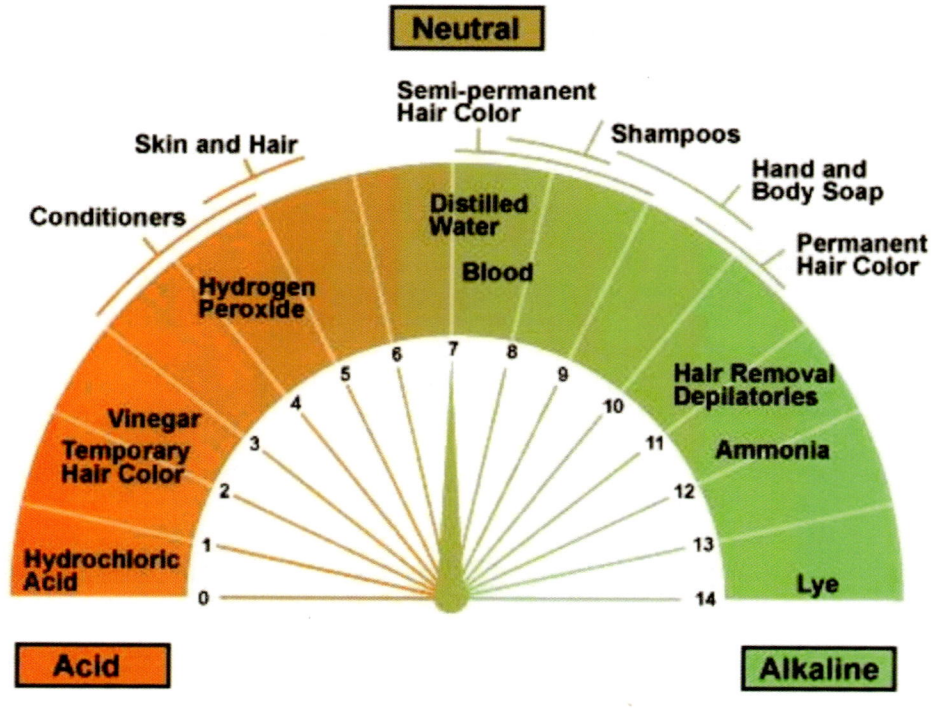

# Alkalinity

**Alkalinity** is a measure of the capacity of water to neutralize acids.

Alkaline compounds in the water such as bicarbonates (baking soda is one type), carbonates, and hydroxides remove H+ ions and lower the acidity of the water (which means increased pH).

# Alkalinity and pH Indicator

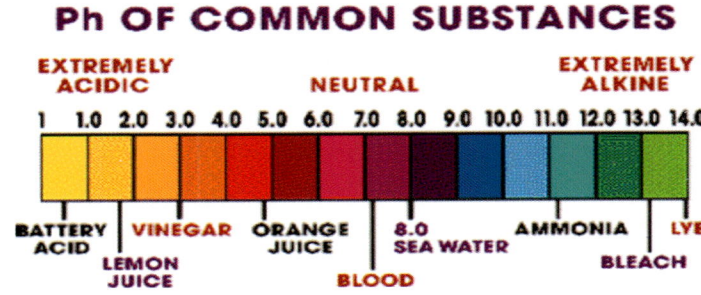

The pH is an indication of the acidity of a substance. It is determined by the number of free <u>hydrogen</u> ions (H+) in a substance.

Water consists of hydrogen ions (H+) and hydroxide ions (OH-).

While more Alkaline is the water more corrosive.

## Salinity Level

The salinity of the oceans changes slightly from around 32ppt (3.2%) to 40ppt (4.0%). Low salinity is found in cold seas, particularly during the summer season when the ice melts. High salinity is found in the ocean 'deserts' in a band coinciding with the continental deserts.

Due to cool dry air descending and warming up, these desert zones have very little rainfall and high evaporation.

The Red Sea located in the desert region but almost completely closed, shows the highest salinity of all (40ppt) – (pH= 8.2 - 8.8 ).

# Fresh Water & Seawater

Seawater <u>pH</u> is typically limited to a range between 7.5 and 8.4.

The average pH in freshwater is 7.0.

**Ph OF COMMON SUBSTANCES**

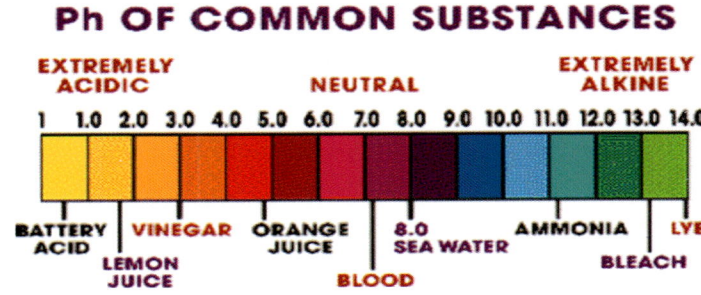

# pH and Acidity

As plants take in CO2 for photosynthesis in aquatic ecosystems, pH values (and **alkalinity**) rise.

Aquatic animals produce the opposite effect -- as animals take in O2 and give off CO2, the pH (and **acidity**) is lowered.

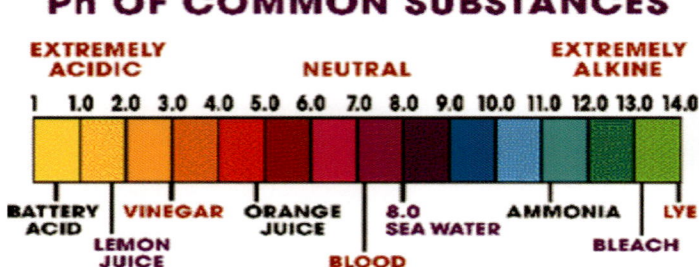

# Pollutants

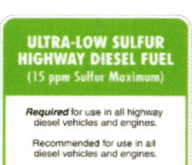

Sulfur dioxide (So2), which is the gaseous product of the combustion of fuels that contain sulfur such as coal, diesel fuel, gasoline, and natural gas, has been identified as one of the most important air pollutants which contribute to the corrosion of metals.

Other corrosion promoters are nitrogen oxides (**No**x), which are also products of combustion. A major source of **NO**x is the fumes from combustion engines. Sulfur dioxide, **NO**x, and airborne aerosol particles can react with moisture and UV light to form new chemicals that can be transported as aerosols.

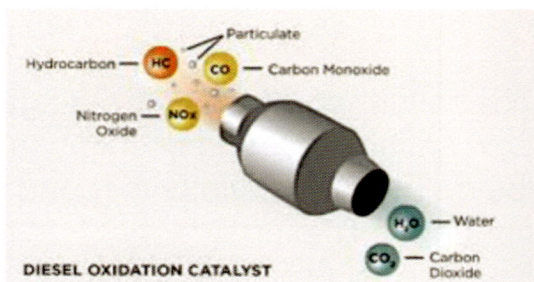

# Corrosion Vs Temperature

For each ten-degree increase in ambient temperature, the corrosion activity can be double.

As the ambient temperature drops during the evening, metallic surfaces tend to remain warmer than the humid air surrounding them.  As the temperature begins to rise in the surrounding air, the lagging temperature of the metal structures will tend to make them act as condensers, maintaining a film of moisture on their surfaces.

## Corrosion on Steel and Cooper

In general, corrosion rates increase with increasing temperature, however for many materials, such as steels, where the oxygen content of the water directly affects the corrosion rate, the effect of temperature is minimal.

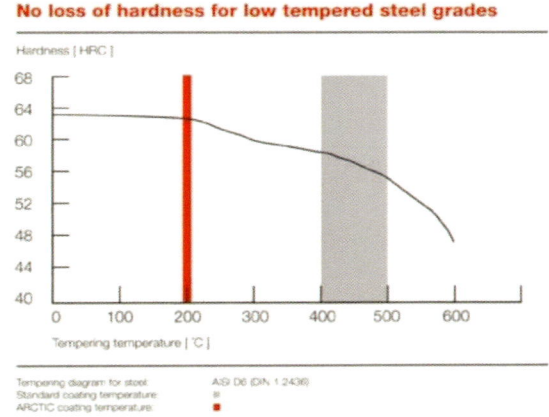

## Temperature Vs Corrosion

The solubility of oxygen is decreased with increasing temperatures and the two effects counteract each other. Steels and copper alloys are particularly insensitive to temperature effects in normal marine immersion.

Many stainless steels have what is essentially a "critical pitting temperature" in seawater that is in the range of temperatures experienced in natural seawater. In cold waters, they do not pit but in warmer waters they are susceptible
Source: "Corrosion Control" NAVFAC MO-307 September 1992.

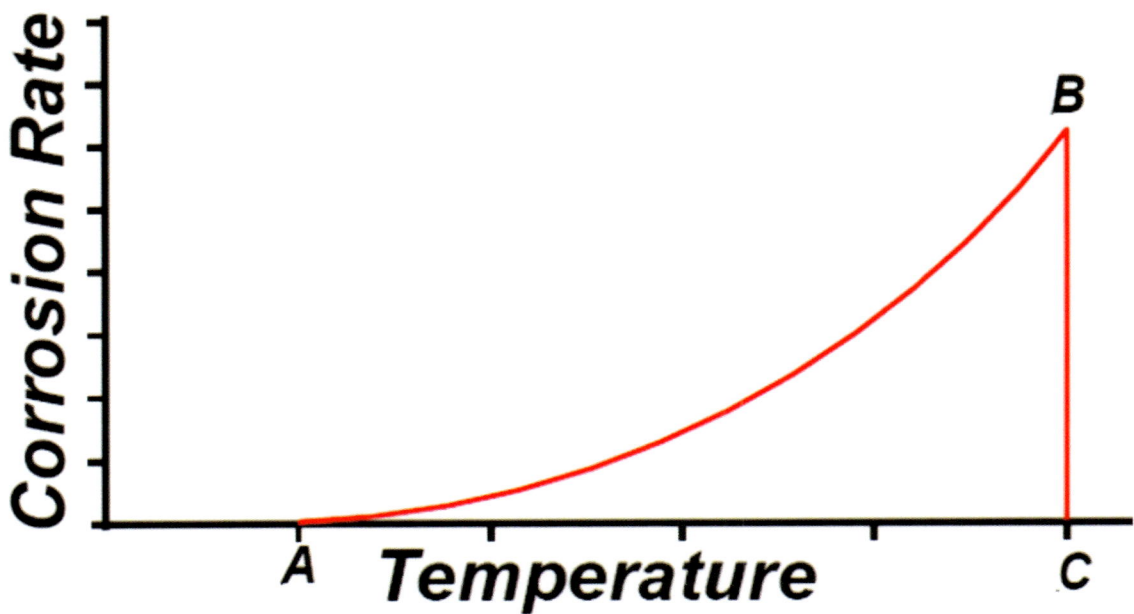

## How The Corrosion is Affected by the Current

### TOPICS

### Video Chapter 6 EP 1: How the Corrosion increase due to a bad Bonding and Grounding

In this video, we will analyze how a wrong grounding system accelerates the corrosion process in a boat and how the problem can be fixed following basic and simple recommendations. This is a great tool for Marine Engineers.

Scan this code to see the highlight video

Follow me

# Factors That Affect the Corrosion

It is clear that the corrosion process of metallic elements on the boat depends on whether the metal/alloy is in direct contact with the sea water or if the metal is not in direct contact with the salt water.

In the first case (Direct Contact) the following factors can affect the corrosion process:

- Acidic and Alkalinity level of the water
- Stray Currents
- Cathodic protection of the metal
- Permanent contact between different metals and Salt water (Galvanism)
- Bonding system quality

In the second case (Indirect Contact) the following factors can affect the corrosion process:

- The humidity level of the environment
- Stray Currents
- Pollutants level on the air
- Bonding system quality
- Coating protection quality
- Inappropriate selection of materials

# What Type of Current Cause Corrosion?

Both types of current (AC and DC) can accelerate the corrosion process.
The more common causes of AC/DC stray-current corrosion are:

- Extension cords dropped into the bilge
- Improper wiring on the boat
- Improper wiring on the dock
- Reverse polarity
- Defective Battery chargers
- Defective Galvanic Isolator
- Neutral and Ground tied on-board
- Shorts to ground
- Inappropriate selection of materials
- Non-Ignition protected devices such as: Starters, Alternators, Ignition Coils and Battery Chargers

# Stray Current

As we mentioned previously the stray currents are the major concern in the corrosion process.

Stray current refers to the electricity flow via buildings, ground, or equipment due to electrical supply system imbalances or wiring flaws.

If an electrical system is not installed or maintained properly, the current can disperse directly to the ground through the equipment or building itself through the grounding conductor. If the system is unproperly grounded into the bonding system, those currents will be stored in the metallic housings of each device in form of static current. The static current will be in constant movement between the dissimilar metals of each device accelerating the Galvanic corrosion between each couple of metals. The more noble metal will attack the less noble eroding the metal progressively.

# Why a Grounding System ?

Any electrical device has a power unit. For example, the Dishwasher has an AC induction motor. When the motor is running, doesn't matter if is ignition protected, some electrons will migrate to the appliance case producing static electricity. If the dishwasher metallic case is not grounded, the static electricity will be there until someone touches the appliance; in that moment the store ed current will be drained through the person's body.

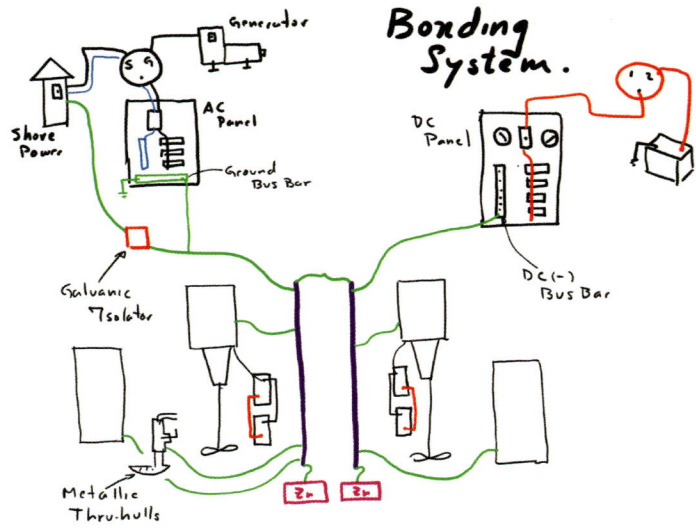

If nobody touches the unit then the static electricity will accelerate the corrosion process in between the dissimilar metals of the unit. Copper rivets will corrode aluminum cases etc, etc.

All the units should be grounded properly, then the grounding conductor should be connected to the external Copper bar buried in the back yard of your home. or to the sacrificial anodes located in the transom of your boat. This is the function of the grounding system in your home and your boat.

Mr. Lopez
Classes.com
For Marine Engineers

# Sources of DC Stray Current

The most common sources of DC stray current are:

- Faulty wiring insulation
- improper wiring
- Poor grounds
- Shorts to ground
- Equipment leaks
- Faulty alternators
- Faulty Battery Chargers
- Faulty Galvanic Isolator

# Stray Current Sources

Stray currents which cause corrosion may originate from **Direct-current** distribution lines, substations, or street railway systems.

**Alternating currents** may cause corrosion especially if the grounding system is defective.

The corrosion resulting from stray currents (**external sources**) is similar to that from galvanic cells (**which generate their own current**) but different remedial measures may be indicated.

However, stray current strengths may be much higher than those produced by galvanic cells and, as a consequence, corrosion may be much more rapid.

# Stray Voltage

**Stray voltage** is the occurrence of electrical potential between two objects that ideally should not have any voltage difference between them.

# Stray Current Corrosion

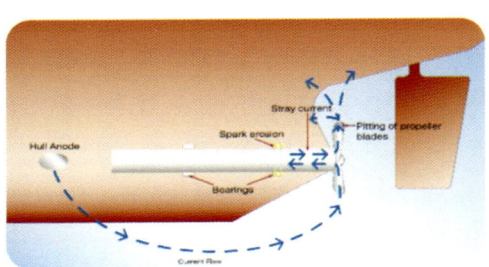

Stray current corrosion or Stray current electrolysis refers to corrosion resulting from stray current flowing through paths other than the intended circuit.

Mr. Lopez
Classes.com
For Marine Engineers

## Video Chapter 6 EP 2: How a Reverse Polarity and a Wrong Wiring Accelerate the Corrosion Process

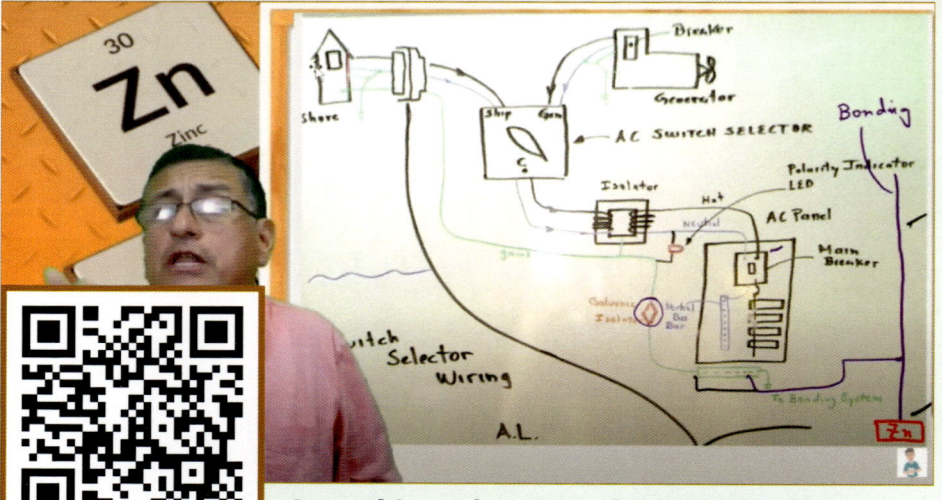

In this video, you will learn how a reverse polarity condition can destroy a boat. The study is accomplished of wiring diagrams and schematics. Enjoy it

Scan this code to see the highlight video

Follow me

# Factors that affect the Corrosion Process

Definitely, it is clear that there are six main factors that accelerate the corrosion process:

1.- Permanent leaks of current
2.- Stored static electricity
3.- Bad Bonding and Grounding Systems
4.- Improper selection of materials
5.- Absence or inappropriate installation of protective devices such as the Galvanic Isolator
6.- A Reverse polarity condition

   *1.- Permanent leaks of current.* There are AC and DC leaks of currents, both of them contribute to the corrosion process. Those leaks can be classified into Intermittent and Permanents.

Sporadic Current leaks occur by bad wiring. Unwanted open wires are the most common source. When those wires touch accidentally metallic surfaces, a short to ground occurs, then if the unit is not Bonded to the system a static amount of electricity will be stored in the metallic case of the unit, starting a slow process of oxidation in between different couples of dissimilar metals.

*Sources:* Engines and Generator harnesses are a common source of current leaks. The excessive engine vibration and high temperatures, melt and break some wires producing intermittent and permanent leaks of current.

Loose terminals, corroded terminals, and bad crimping terminals are other sources of current leaks.

*2.- Stored static electricity*: It is a silent source of small DC currents. Due to the absence of good grounding conductors, those small currents, in presence of environmental humidity accelerate the corrosion process in between dissimilar metals. All the metallic elements in a boat should be bonded into the main central bonding conductor Does no matter if the element is located over or below the waterline level or if the element is or is not in permanent contact with the seawater.

*Sources:* The use of non-ignition-protected devices is the main source of static currents. The second source is bad wiring to the main grounding conductor.

*3.- Bad Bonding and Grounding Systems*. The bonding network and the grounding connection are the heart of the system. As you can see in the annex video-Clip, the life of your vessel depends on the quality of the bonding connections and the proper connection to sacrificial anodes. It is too simple if you follow the recommendations to select the appropriate sacrificial anodes according to the hull material and type of water (See the annex Poster)

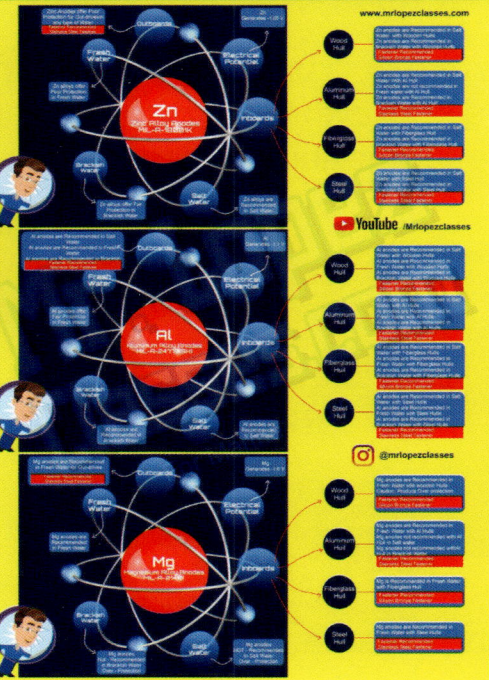

**Scan this code
to watch the video.**

**Scan this code
to buy this poster**

 **4.- Improper selection of materials:** This should be the fundamental concept for a Marine or Naval Engineer. The selection of marine grade materials is the basis for a great design. Under any conditions use residential material in a boat project ; you can

ruin your investment for a simple "Brass fitting".

In the Annex table you will find the most recommended Ferrous and Non-Ferrous alloys recommended for marine applications.

 **5.- Absence or inappropriate installation of protective devices such as the Galvanic Isolator:** The only way to protect your boat from neighbor boat current leaks and also from the intrusion of bad currents through the ground conductor is with a good Galvanic Isolator properly installed and periodically tested according to the recommendation in the Annex video-Clip.

Some technicians and boat owners confuse the Galvanic Isolator with the Isolator Transformer. To prevent corrosion issues in your boat you should install a Galvanic Isolator.

The Isolator transformer is used to clean the input AC signal (Eliminates Noise and Interference in the AC wave) to the main AC panel. To clarify the difference between those units please see the annex Video-Clip)

**Scan this code
to watch the video**

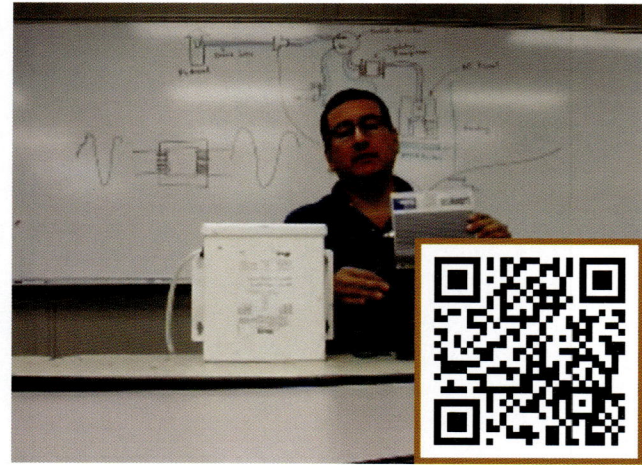

**Scan this code
to watch the video**

  6.- *A Reverse polarity condition:* This is another concept difficult to relate with the corrosion process for Engineers and Technicians.

The reverse polarity indicator is a LED light with two legs. One leg with a resistor connected to the Ground busbar and the other one connected to the Neutral bus bar at the AC panel. In normal conditions with good panel wiring, this light should be Off because no current should be flowing to the Neutral conductor. Remember that the number recommendation for good boat wiring is always to keep the Ground and Neutrals bus bars isolated from each other. ( They should be connected together ONLY at the power source).

It is important to refresh also that both, Negative DC bus bar and the Ground AC bus bar should be connected together at the main bonding conductor.

Reverse Polarity Vs Corrosion.

Now, suppose that for some reason somebody switches the AC HOT line with the Neutral line before the main breaker at the AC panel, then right now the Neutral bus bar will be a HOT bus bar and of course, the Hot bar will be neutral. And now suppose that somebody connects accidentally the Neutral with Ground (A common residential error).

Well, this is the worse scenery, in this moment the AC ground bus bar and also the Neutral bus bar are HOT bars, then all the metallic elements previously bonded into the main grounding conductor will be receiving a permanent flow of AC current, accelerating the corrosion process in those elements. If this scenario is not fixed quickly some metallic through hulls will be eroded in a couple  of days with catastrophic consequences.

In the next Video Clip, you can learn how a Reverse Polarity condition accelerates the corrosion process in a boat

**Scan this code to watch the video**

# Tools Recommended for Testing

This is the list of specialized tools recommended to perform a proper corrosion survey:
- A Digital Multimeter
- An Ideal Sure-Test  meter (AC Circuit Analizer)
- A Silver/Silver Chloride reference cell
- A  portable Galvanic Isolator with 30 Amps plugs
- A Shorepower cord with split lines to insert a clamp meter
- Assorted  suspended  fish anodes (Zn , Al and Mg)

# Permanent Monitoring Tools

Additional to the regular Amp-meters, Digital Amp-Meters calibrated in milliamps should be installed in both, AC and DC panels. Those meters allow you to identify the flow of small currents even if the breakers and main interrupters are Off. This is in my opinion the best tool to identify if small parasite currents are flowing through the circuits.

Currents over 1 milliamp should  are  considered significant and should be analyzed in details

**Video Chapter 6 EP 3: How to Reduce The Corrosion Process & Identifying Leaks of Current**

In this video, you will learn different techniques to identify small leaks of DC current. Using a panel board we will learn how to measure those leaks. Enjoy it

**Scan this code to see the highlight video**

**Follow me**

# Leakage Current Testing Procedure (AC Current)

In normal conditions, none type current should be flowing through the grounding conductor. however, the ground line could be affected by factors such as:

- Reverse Polarity
- Neutral and ground tie-in on board
- Shorts to ground
- Proximity between AC and DC Panels
- A defective:

  ◦ Galvanic Isolator
  ◦ Isolator Transformer
  ◦ Battery Charger
  ◦ Inverter
  ◦ Wiring on AC Panel

# Leakage Current Testing Procedure (DC Current)

We can determine a leakage current problem with three different types of tests.

With the **voltage** measurement test, we can find the equipment that is leaking current constantly.

With the **amperage** measurement test, we can determine the amount of current that is flowing through the leakage path.

With the **resistance** measurement test, we can determine how much the circuit is restricting the flow of current.

# Voltage Leakage Test

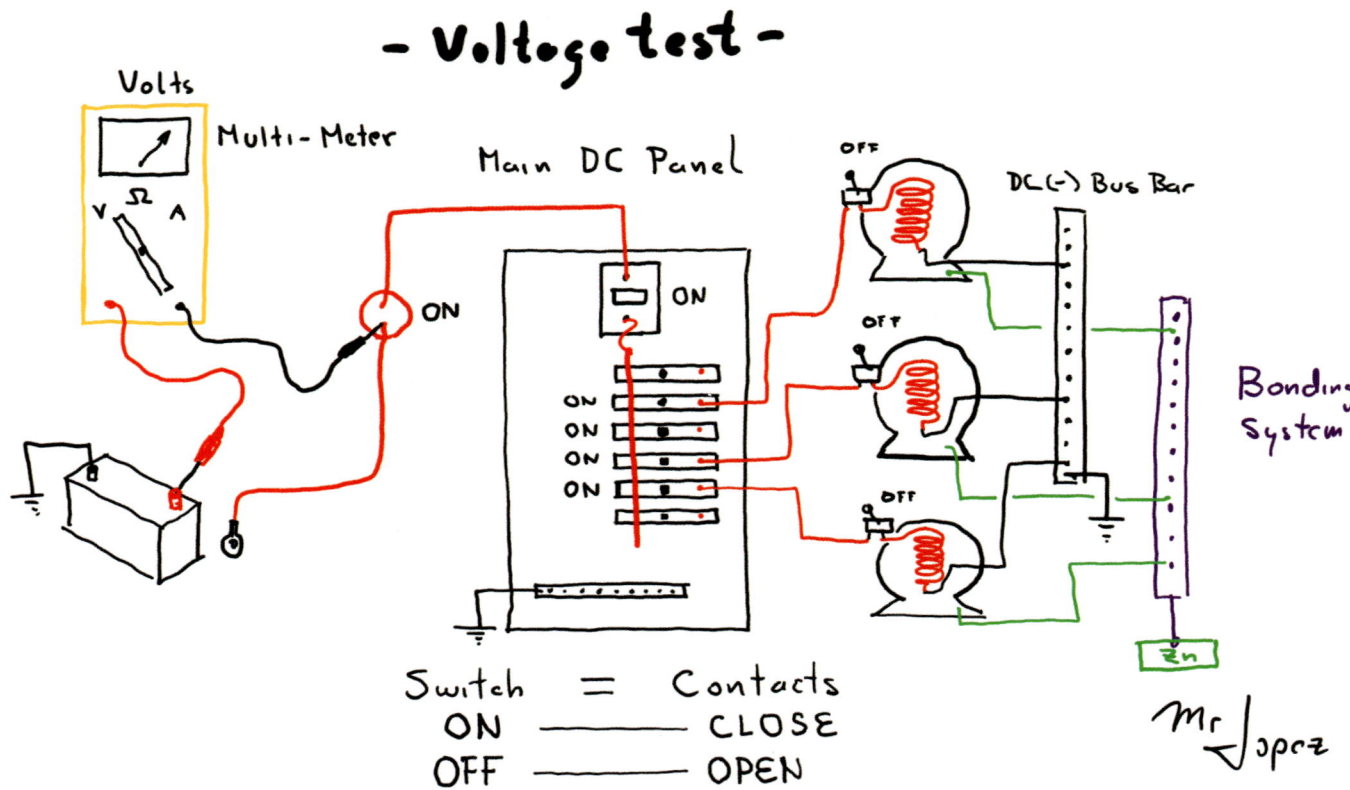

Since all equipment switches are open (Off Position) there should be no complete circuit back to the negative terminal of the battery, and the voltmeter should read zero volts.

If the reading is 12 volts, then either some appliance is still on or there is a leakage path.

Verify that all the equipment is in of. If the reading stays at 12 volts then the leakage path exists.

## Further Analysis (Volt)

If we analyze the "Voltage Test Drawing", each equipment has its own switch and those interrupters are in "'off" position. In other words, its internal contacts are open. If there is 12 V on the multimeter it is an indication that through one of those interrupters is passing a small current (Not enough to start the unit) closing the circuit to show 12V at the meter display.

The procedure to identify what is the unit with the issue is disconnecting the power wire entering in each unit. If the voltage reading disappears, then this is the unit in trouble.

## Current Leakage Test (Amp)

Set the meter to read DC amps. Fix the red test probe to the positive battery terminal and the black one to the disconnected positive battery cable.

If the leaking current is less than 1 milliamp is not considered significant.

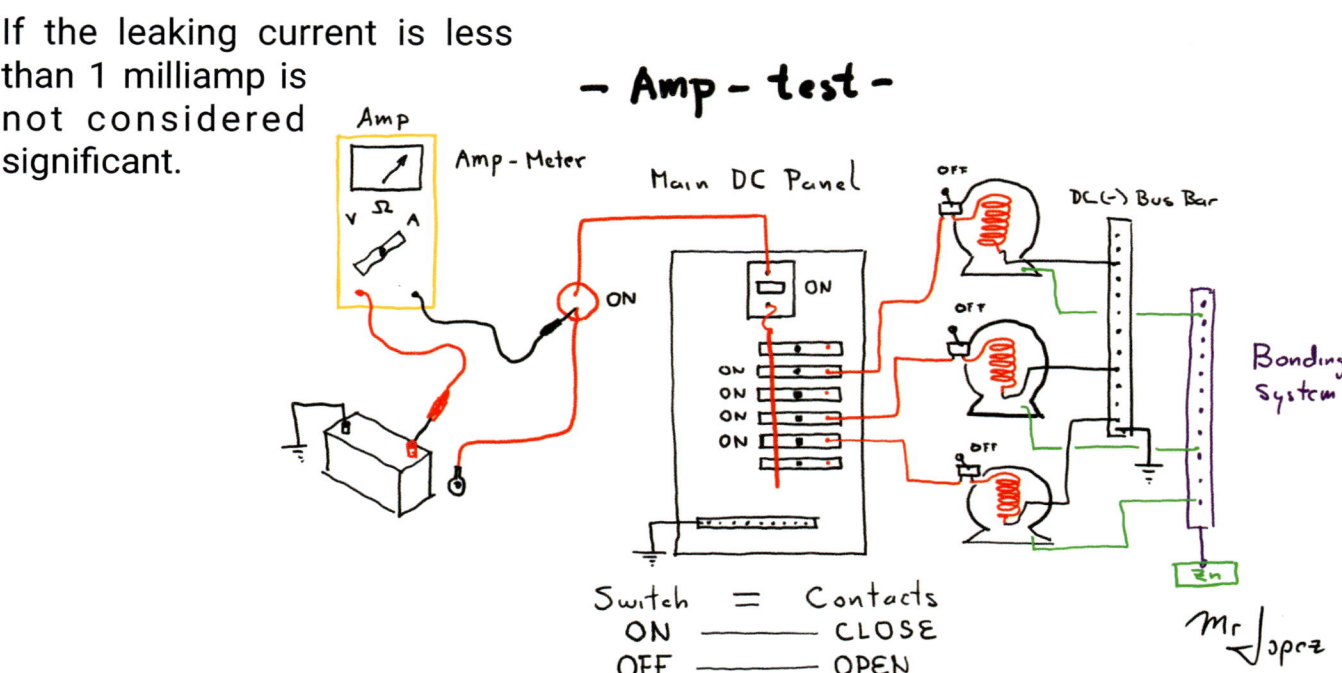

## Further Analysis (Amp)

The further analysis is similar. In this case, the Amp-meter is reading more than 0.001 Amp. This is an indication that through any of the units is flowing a small current. By removing the power into each unit you should check if the reading goes down to zero, indicating that this is the unit in trouble.

## Resistance Leakage Test (Ohm)

Due then we don't know how much current is flowing through the leakage path; that depends on how much resistance there is in the path.

With the positive battery cable disconnected and all the equipment switches set to the Off position, set the multimeter to read "Ohms'

A Resistance reading above 1.2 Ohms means you can begin to isolate the leakage path.

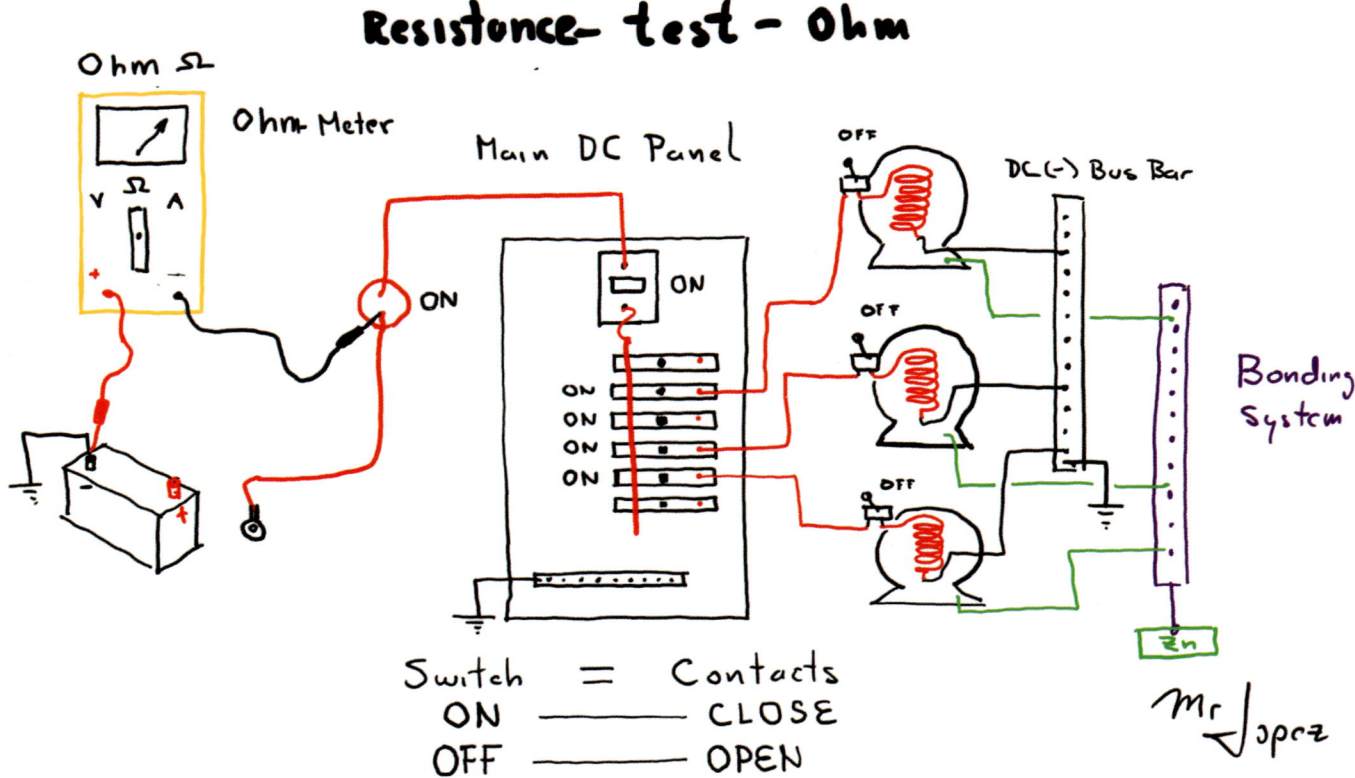

## Further Analysis (Ohm)

In this case, the Ohm-meter is reading more than 1.2 Ohms. When the current increases then the circuit resistance increases. This is an indication that through any of the units is flowing a small current increases the resistance in the system. By removing the power into each unit you should check if the reading goes down to zero, indicating that this is the unit in trouble.

# Common Ground Point

The only thing that the AC and DC panels have in common is the ground point or bonding system.

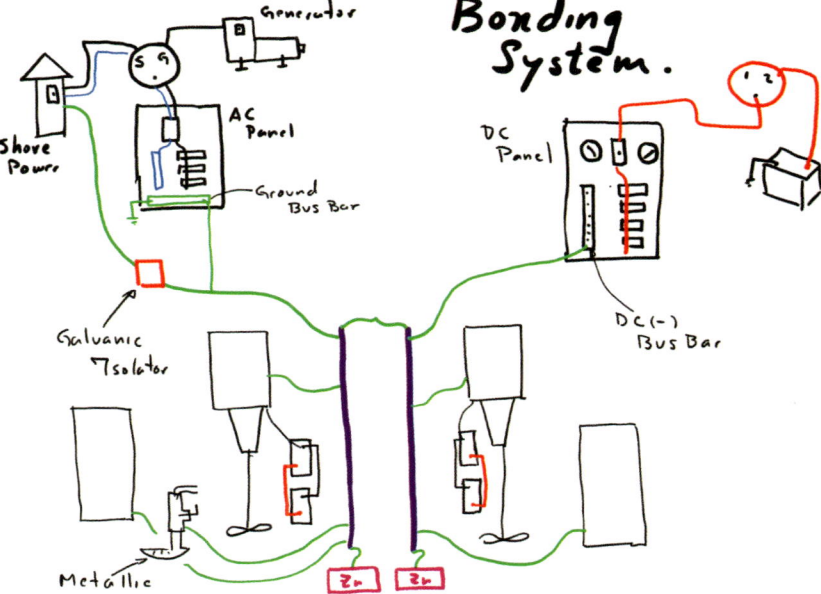

Marine electrical systems are grounded through a ground bus bar that is connected to the bonding system.

# Neutral and Ground

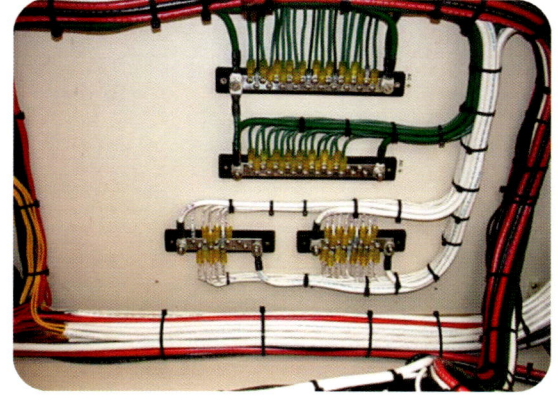

Do you remember what is the only common element between the AC and DC panels?
It is correct. The grounding, both the DC negative bus bar and the AC ground bus bar shall be connected together and then connected to the bonding system.
Now we are going to refresh the concepts studied in chapter 12 of my book of "Marine Electricity" about Neutral and Ground connections and Reverse Polarity.

# Neutral & Ground Tie-In

With 120 VAC systems, the green and white get tied together at any <u>AC power source,</u> but only at a source of AC power.
Next, consider potential power sources used for boats:

- Shore power delivered from the dock
- Generators
- Inverters

# Neutral & Ground Tie-In

The shore power neutral is grounded through the shore power cable and shall not be grounded on board the boat.

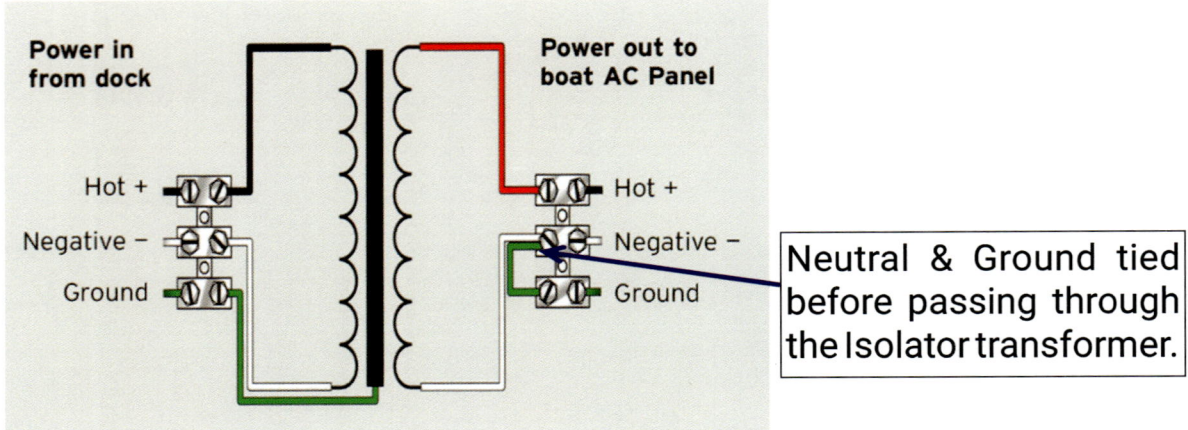

Neutral & Ground tied before passing through the Isolator transformer.

# The Galvanic Cell

What happens if your boat is properly fitted out with zincs, but your neighbor hasn't been as fastidious and has no zincs?
As soon as the other boat plugs into shore power, a big galvanic cell is created.

The zincs on your aluminum stern drive become the anode, while the bronze propeller on the other boat becomes the cathode.
The metallic path for electrons is from our zincs through the wires to the common ground point.

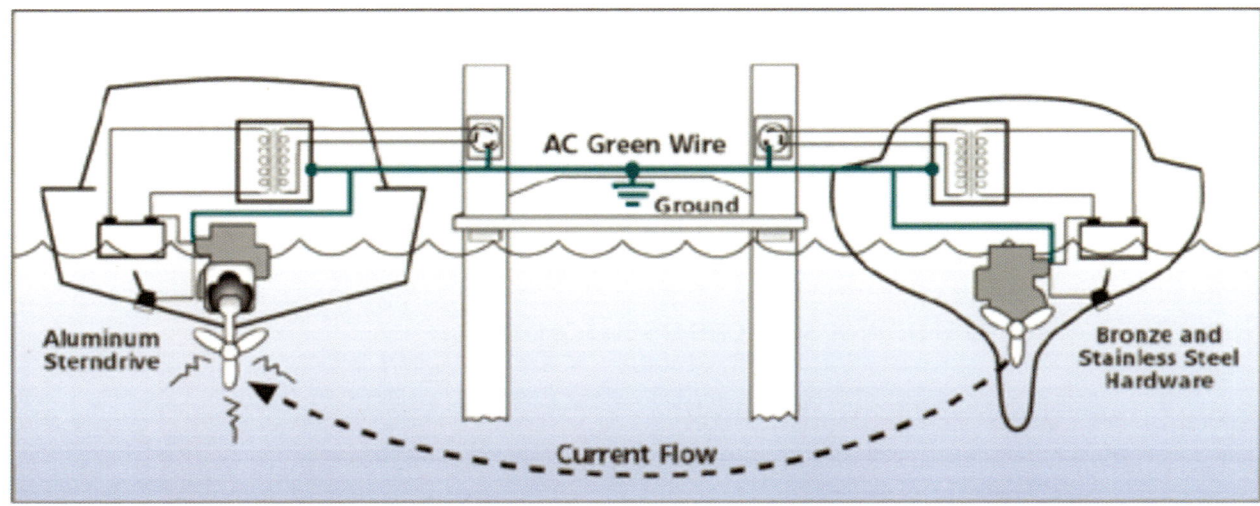

## Shore Lines & Corrosion

These devices are exposed to water and overtime suffer from corrosion and general wear.

High resistance caused by corroded, bent, or worn connectors results in high resistance which causes overheating, which further amplifies the power drop.

## Shore Lines

Make sure you turn off the breakers at your boat before the pedestal! Reverse the order when attaching the cord and you are all set!

Not only do you risk getting electrocuted, but disconnecting an energized connector damages the contacts. Also, consider what happens if you drop the energized cord in the drink!

## Main Breakers

The circuit breakers on the pedestal protect only the dock wiring. Your main circuit breaker protects your boat systems.

Normally the section of wiring and connectors between your main panel and the dock is unprotected.

# Analysis of Tools Recommended for Testing

Now we are going to study some of the best tools used to analyze corrosion issues.

## Galvanic Isolator

A galvanic isolator is a device used to block low voltage DC currents coming on board your boat on the shore power ground wire. These currents could cause corrosion to your underwater metals; through-hulls, propeller, shaft, etc.

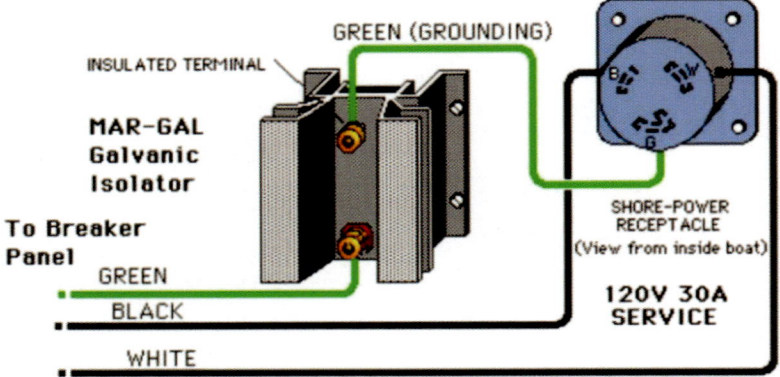

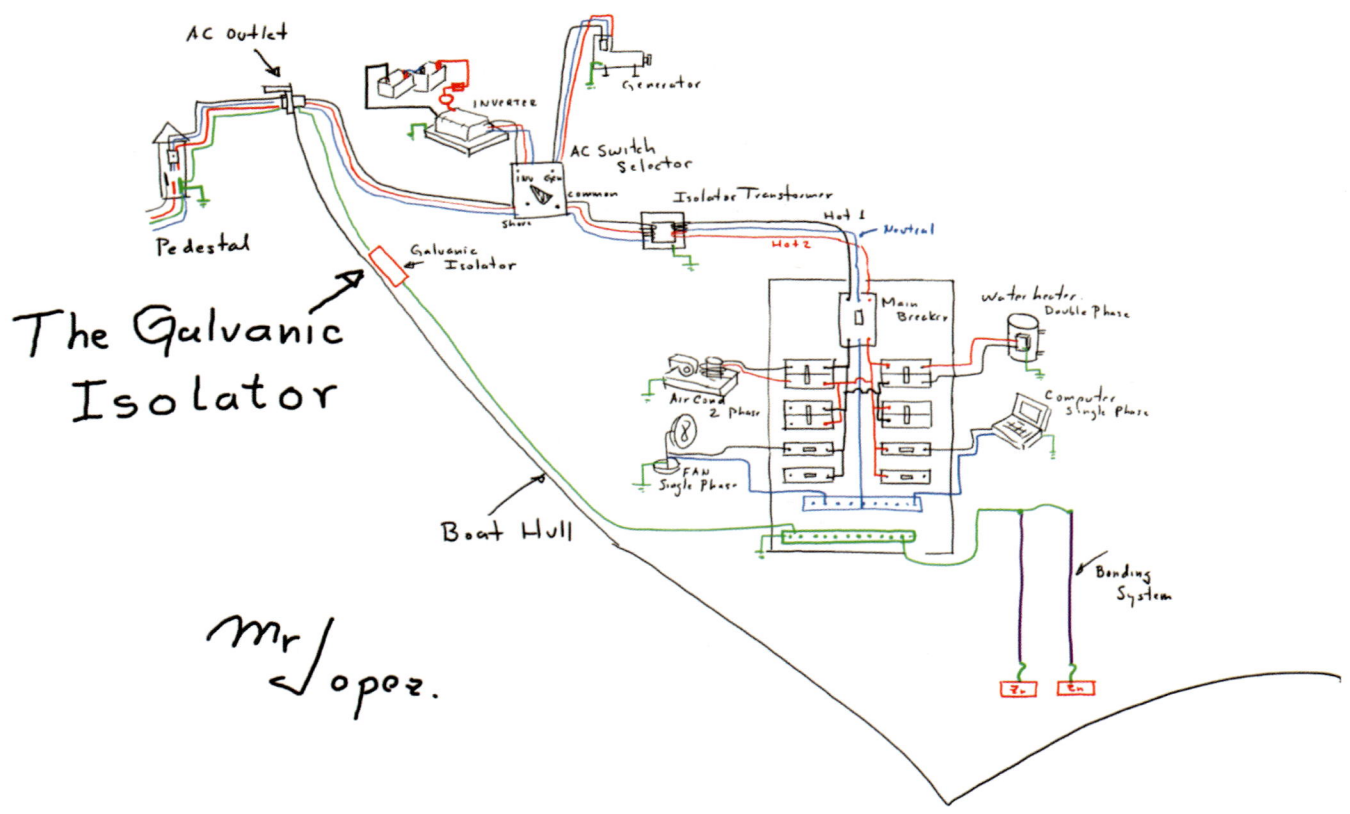

The Galvanic Isolator

Mr. Lopez.

Mr. Lopez Classes.com
For Marine Engineers

## Galvanic Isolator & Corrosion

The Galvanic Isolator blocks electrolysis currents from flowing in the ground conductor of your shore power hookup. It provides approximately 1.2 volts of isolation to isolate electrolytic voltages from the dock but yet passes safety currents to the ground in the event of a short circuit, or power leakage on your boat.

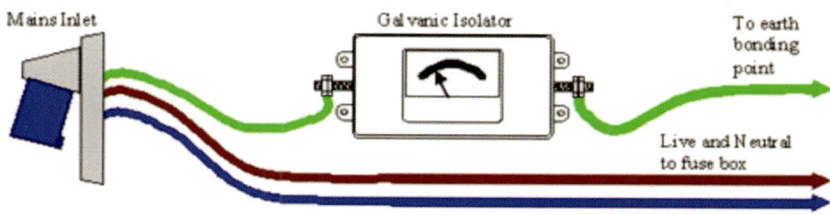

## A Defective Galvanic Isolator

If the ground line coming from the shore power cord is contaminated and if the galvanic isolator is defective, then the low voltage DC current entering your boat will affect the boat bonding system, contaminating both, the AC ground bus bar and the DC negative bus bar.

The Galvanic Isolator will block those currents protecting the boat wiring .

The Galvanic Isolator should be inspected once per year according to the procedure recommended in the video.

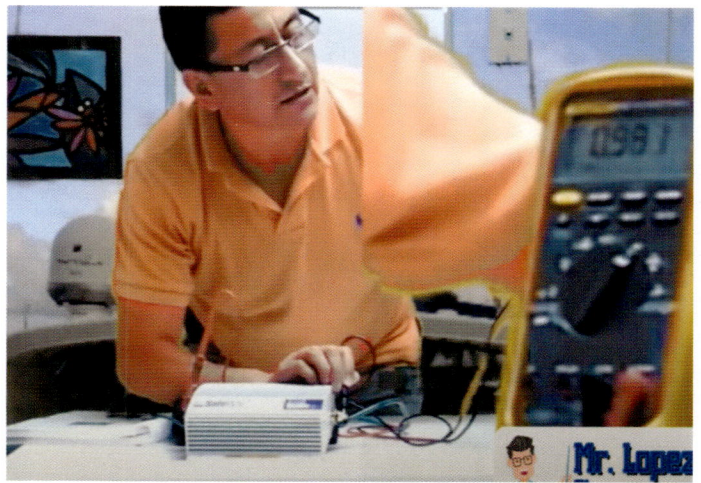

Normally after severe storms, the Galvanic Isolator is affected severely. In those cases, it should be replaced for a new one with the same capacity in Amps.

Remember that the capacity of the Galvanic Isolator is according to the main breaker capacity.

## Video Chapter 6 EP 4: The Silver Chloride Electrode Test

In this Video you will learn the procedure to use the silver chloride electrode test, to verify if your boat is properly protected against corrosion. This video is a useful tool for Boat Owners, Inspectors, and Marine Engineers. Enjoy it

**Scan this code to see the highlight video**

**Follow me**

# Silver Chloride Test Procedure

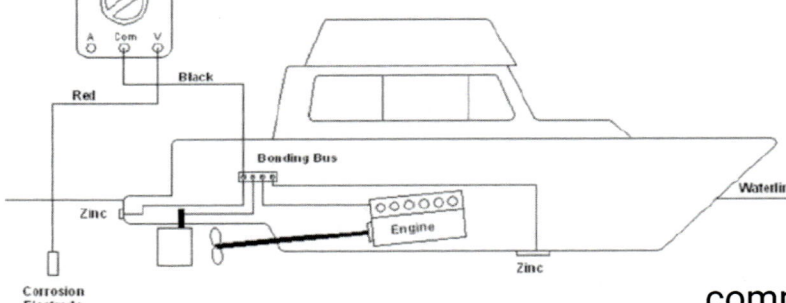

The tools you'll need for the job include a high-quality digital multimeter and a reference electrode.

## The Electrode

Reference electrodes are often called "half cells" because they contain a metal and a metal compound. Popular types are Copper-Copper Sulfate and Silver-Silver Chloride.

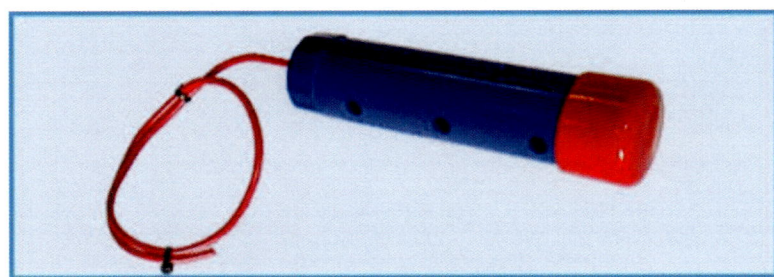

# The Procedure

Set the multimeter function to DC volts. Connect the reference electrode to the volts input jack and place the electrode in the water. Best results are obtained when the electrode is located away from the anode.

Connect the multimeter common jack to a probe that will be used to contact each piece of underwater metal.

Touch the common probe to each underwater metal fitting and record the millivolt value as displayed on the meter.

# The Readings

If all underwater metal fittings are connected together with a "bonding" system, then all readings should be identical.

| Meter Reading (Volt DC) | Cathodic (Corrosion) Protection Status |
|---|---|
| More Positive than (-0.7) | Non Protection |
| Between (-0.7) and (-0.8) | Inadequate Protection |
| Between (-0.8) & (-0.95) | Adequately Protected |
| Between (-0.95)& (-1.05) | Adequately Protected- High Range Not recommended for Aluminum Hulls |
| More Negative than (-1.05) | Over Protected |

# Overprotection

Overprotection can cause paint to peel from a metal hull or chemical damage to a wooden hull.

# Battery Chargers

One of the more common causes of stray current corrosion is the use of automotive battery chargers.
Marine battery chargers are designed to have separate AC and DC ground points.

All of the currents brought to the charger from the AC source through the hot wire are returned to the source through the neutral wire.

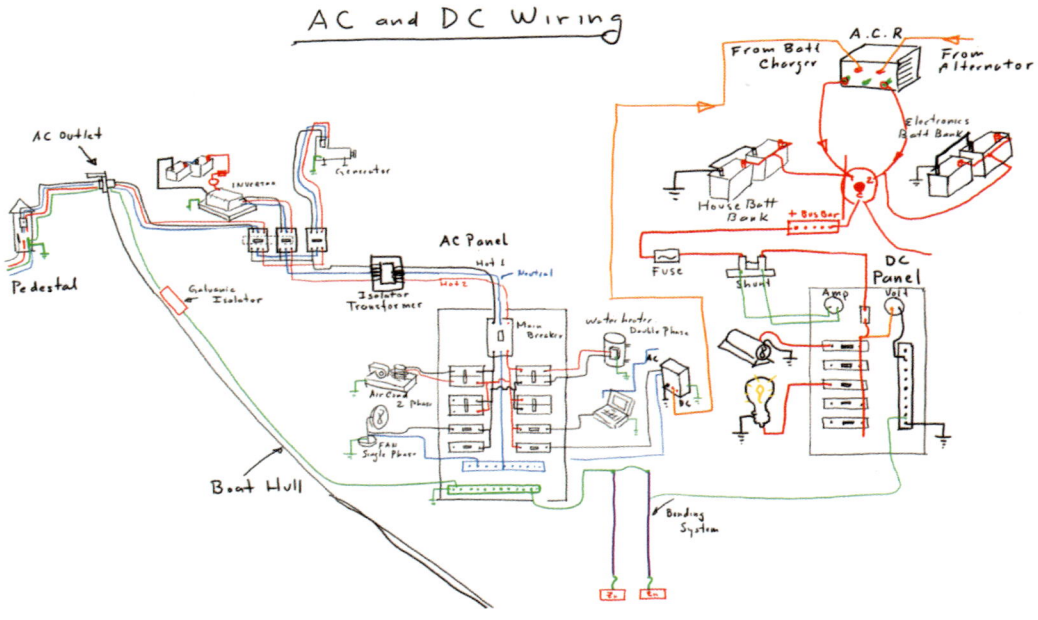

# Universal Sacrificial Anodes

A good tool to verify if your boat is under-protected is the grouper zinc anode.
There are Sacrificial anodes in different materials such as Zn, Al, and Mg.
During the Silver/Silver Chlorine test method, adding portable sacrificial anodes by means of the alligator terminals directly to the main bonding conductor, you can verify if the readings increase or decrease, indicating over or under galvanic protection.

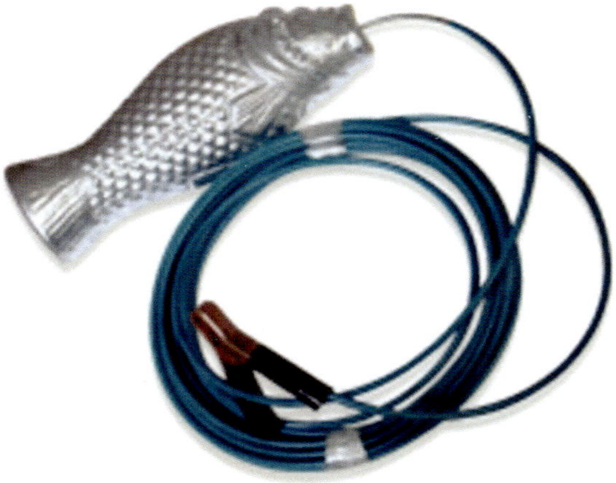

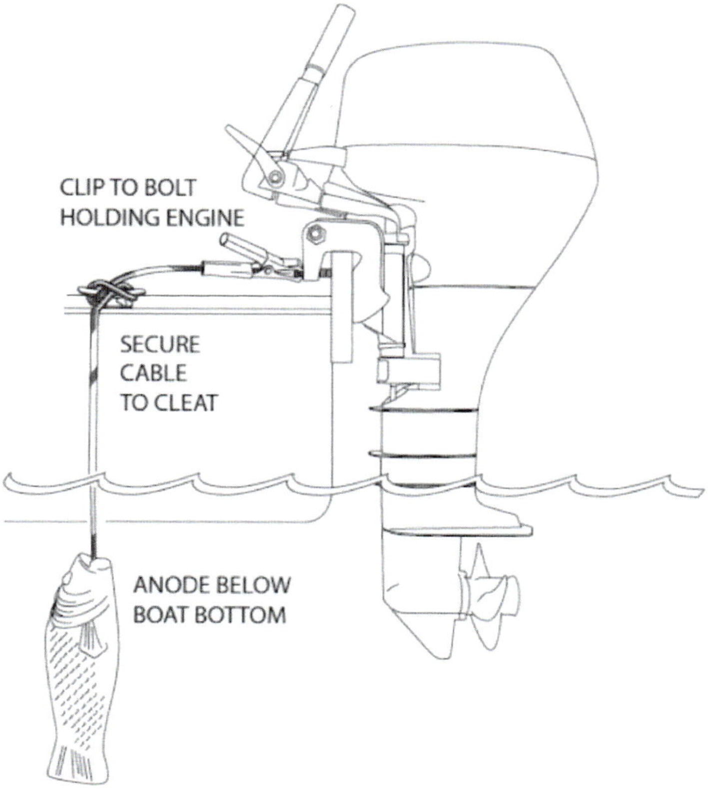

CLIP TO BOLT
HOLDING ENGINE

SECURE
CABLE
TO CLEAT

ANODE BELOW
BOAT BOTTOM

ZNGUY 3

# The Ideal Sure Test

The Ideal Sure Test  is a comprehensive branch circuit analyzer that allows you to "look behind walls" to find wiring problems saving you time and money while letting you get the job done with safety and confidence.   its advanced processing software gives it the capability to quickly identify bootleg grounds at any receptacle.

SureTest helps you to verify proper installation and periodically test branch circuits to confirm integrity of wiring and protection devices to ensure continued protection from shock hazards.

Many residential fires are the result of a high resistance point leading to arcing within a branch circuit. Loose and corroded connections, bad splices, faulty cords and defective devices are common causes of many structural fires.

Sure Test Conducts testing without disturbing sensitive loads or tripping breakers

Identifies proper wiring polarity in 3 wire receptacles

Identifies false (bootleg) grounds, without opening up the outlet

Indicates voltage drop and impedance on live hot, neutral and ground conductors

Allows you to zero in on the problem conductor and gives you clues as to the nature of the issue.

SureTest measures and displays individual Line Impedance giving the electrician/inspector valuable insight as to whether a particular conductor is the problem due perhaps to faulty splicing, overly long conductor runs, or undersized wire gauge.

**Scan this code to watch the video.**

## Types of Corrosion

### TOPICS

### Video Chapter 7 EP 1: Types of Corrosion

This Video is an analysis of the different types of corrosion in boats, depending on the hull material, the type of water, and the bonding and grounding. Enjoy it!!!

Scan this code to see the highlight video

Follow me

## The Sea Water Components

The marine environment and saltwater provide the elements necessary for the corrosion process.

## The Marine Environment

Don't forget that seawater contains minerals, metals, and non-metals. Those components can accelerate the corrosion process because they can produce stable molecules easily.

**The Major Dissolved Constituents of Seawater**

| Ion | $Cl = 19‰$ | Percent |
|---|---|---|
| Cl | 18.980 | 55.05 |
| Br | 0.065 | 0.19 |
| $SO_4$ | 2.649 | 7.68 |
| $HCO_3$ | 0.140 | 0.41 |
| F | 0.001 | 0.00 |
| $H_3BO_3$ | 0.026 | 0.07 |
| Mg | 1.272 | 3.69 |
| Ca | 0.400 | 1.16 |
| Sr | 0.008 | 0.03 |
| K | 0.380 | 1.10 |
| Na | 10.556 | 30.61 |
| Total | 34.477 | 99.99 |

### The Composition of the Atmosphere at Ground Level

| Gas | Concentration by Volume (ppm) | Concentration by Weight (ppm) |
|---|---|---|
| Nitrogen, $N_2$ | 780,840 | 755,220 |
| Oxygen, $O_2$ | 209,460 | 231,400 |
| Argon, Ar | 9340 | 12,880 |
| Carbon Dioxide, $CO_2$ | 320 | 490 |
| Neon, Ne | 18.18 | 12.67 |
| Helium, He | 5.24 | 0.724 |
| Krypton, Kr | 1.14 | 3.30 |
| Xenon, Xe | 0.087 | 0.39 |
| Hydrogen, $H_2$ | 0.55 | 0.038 |
| Nitrous Oxide, $N_2O$ | 0.33 | 0.50 |

Seawater is a better electrical conductor than freshwater due to the presence of metals and minerals, that is why seawater is considered a good electrolyte.
Now we will analyze how a marine environment plus the presence of current leaks is the worse scenario to accelerate the corrosion in a boat.

# Common Types of Corrosion

- Erosion Corrosion = Impingement damage
- Crevice Corrosion = Stagnant solution
- Poultice Corrosion = Dirt and debris accumulate
- Pitting Corrosion = localized form of corrosion by which cavities or "holes
- Galvanic Corrosion = Is an electrochemical process in which one metal corrodes preferentially to another
- Intergranular Corrosion= Corrosion in the neighbor of a welding area.

# Types of Corrosion

### Galvanic Corrosion

Galvanic Corrosion also called Bimetallic Corrosion is a common mode of corrosion failure that occurs when two dissimilar metals are together in a marine environment. The more noble metal (more electronegativity) attacks the less noble (less electronegativity). The less noble metal donates the electrons on its outer orbit producing erosion in its surface.

Galvanic Corrosion is considered a local Form of Corrosion. It is limited to the contact zone in between both metals. The intensity of the process decreases rapidly with increasing distance, even by a few centimeters, from the point of contact between both metals. This decrease is greater when the electrolyte is a poor conductor.

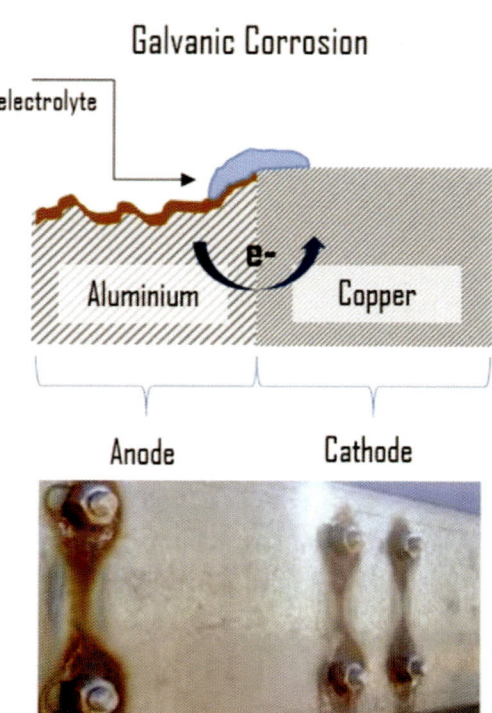

This type of corrosion is so localized because of electrical reasons. The small arc of electricity is produced in the contact area between both metals. The intensity of the electrical arc depends on the potential difference ( Electronegativity) between both metals. Since Galvanic Corrosion often tends to develop at depth. it is not uncommon that galvanic corrosion perforates parts several millimeters thick.

In other words, the intensity of the galvanic corrosion depends on the potential difference between the metals in contact.

# Galvanic Corrosion

After checking the Electronegativities table (Pag 9), we can analyze the reaction between Cu and Al. The Copper with (1.9) is nobler than Aluminum (1.5), then the Al piece will be eroded. In this case, the Al piece is the anode (+), and the Cu piece is the cathode (-)

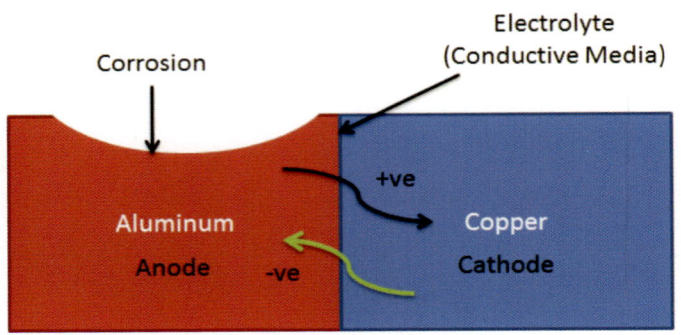

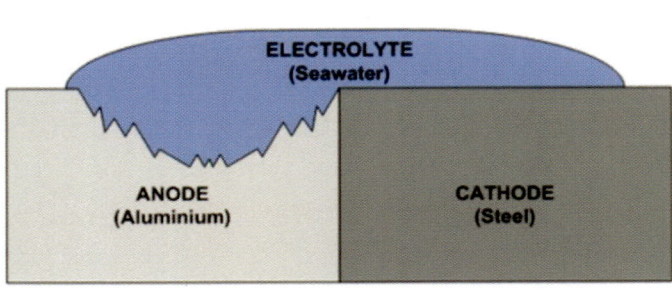

Another great example is the relation between Aluminum and Steel (Iron Alloy). Iron with (1.8) is nobler than Aluminum (1.5), then the Al will be eroded (Sacrificial Anode), while steel will gain electrons workings as a cathode (-)

## Electronegativity Table

Maximum

| | | | | | | | | | | | | | | | | | |
|---|---|---|---|---|---|---|---|---|---|---|---|---|---|---|---|---|---|
| H 2.1 | | | | | | | | | | | | | | | | | |
| Li 1.0 | Be 1.5 | | | | | | | | | | | B 2.0 | C 2.5 | N 3.0 | O 3.5 | F 4.0 |
| Na 0.9 | Mg 1.2 | | | | | | | | | | | Al 1.5 | Si 1.8 | P 2.1 | S 2.5 | Cl 3.0 |
| K 0.8 | Ca 1.0 | Sc 1.3 | Ti 1.5 | V 1.6 | Cr 1.6 | Mn 1.5 | Fe 1.8 | Co 1.9 | NI 1.9 | Cu 1.9 | Zn 1.6 | Ga 1.6 | Ge 1.8 | As 2.1 | Se 2.5 | Br 3.0 |
| Rb 0.8 | Sr 1.0 | Y 1.2 | Zr 1.4 | Nb 1.6 | Mo 1.8 | Tc 1.9 | Ru 2.2 | Rh 2.2 | Pd 2.2 | Ag 1.9 | Cd 1.7 | In 1.7 | Sn 1.8 | Sb 1.9 | Te 2.1 | I 2.5 |
| Cs 0.7 | Ba 0.9 | La⁻ 1.0 - 12 | Hf 1.3 | Ta 1.5 | W 1.7 | Re 1.9 | Os 2.2 | Ir 2.2 | Pt 2.2 | Au 2.4 | Hg 1.9 | Tl 1.8 | Pb 1.9 | Bi 1.9 | Po 2.0 | At 2.2 |
| Fr 0.7 | Ra 0.9 | | | | | | | | | | | | | | | |

Minimum

# Galvanic Corrosion in Sacrificial Anodes

The galvanic cell Fe-Zn is a great example of how a sacrificial anode works. In this case, Fe the nobler metal (1.8) will be protected by the Zn (sacrificial anode) lesser noble (1.6), surrounded by saltwater. A similar situation occurs between Cu (1.9) and Zn (1.6); however, in the galvanic cell between Zn(1.6) and Al(1,5) is evident that Zn is not a good protector for Al in that case a sacrificial anode in Magnesium (1.2) will work better.

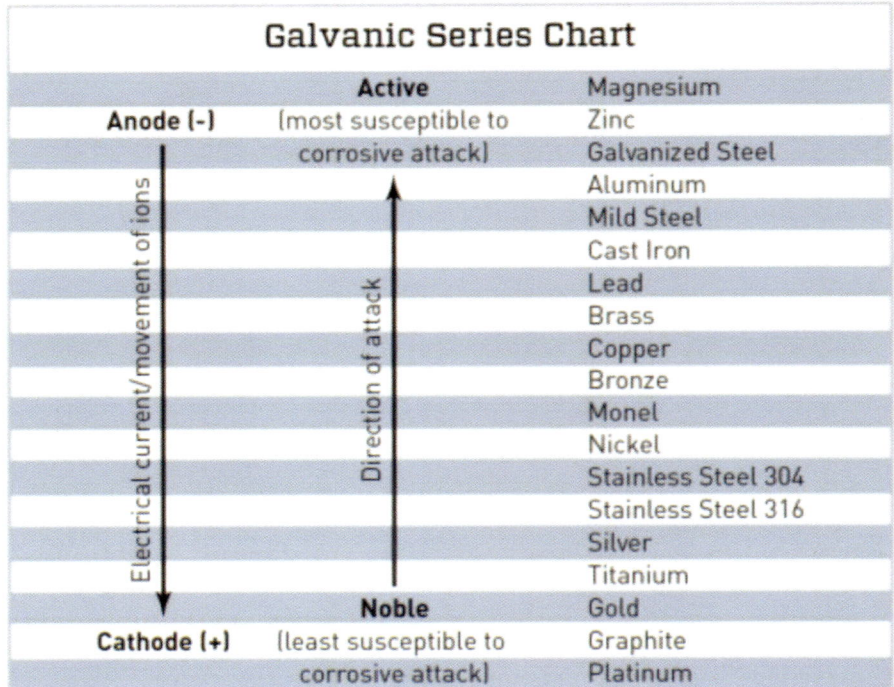

In the galvanic series chart, you can see that Mg is more Anodic than Aluminum, then Mg will be a good protector for Al structures.

The galvanic series chart is a great tool to understand which metal is a better protector(more anodic) than another when talking about marine structure designs.

As a rule, the further apart the metals are on the table, the more they will react in the presence of an electrolyte.

## Galvanic Series Chart

| | | |
|---|---|---|
| | **Active** | Magnesium |
| Anode (-) | (most susceptible to | Zinc |
| | corrosive attack) | Galvanized Steel |
| | | Aluminum |
| | | Mild Steel |
| | | Cast Iron |
| | | Lead |
| | | Brass |
| | | Copper |
| | | Bronze |
| | | Monel |
| | | Nickel |
| | | Stainless Steel 304 |
| | | Stainless Steel 316 |
| | | Silver |
| | | Titanium |
| | **Noble** | Gold |
| Cathode (+) | (least susceptible to | Graphite |
| | corrosive attack) | Platinum |

(left axis: Electrical current/movement of ions; inner axis: Direction of attack)

Mr. Lopez Classes.com
For Marine Engineers

# Galvanic Corrosion

Occurs when two different metals are located together in a corrosive underlineelectrolyte (Salt Water). A galvanic couple forms between the two metals, where one metal becomes the anode and the other the cathode.

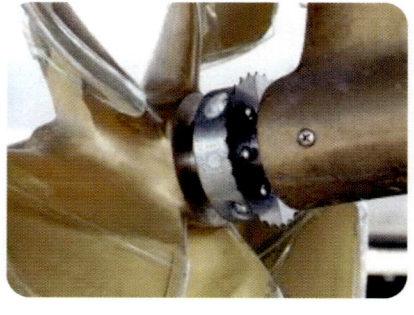

Any water which contains impurities will act as an electrolyte.

The risk of galvanic corrosion is reduced by limiting the variety of materials used during boat construction and by ensuring sacrificial anodes are installed. This is particularly important on Aluminium boats.

The picture below shows the impact of galvanic corrosion caused by cylinder head

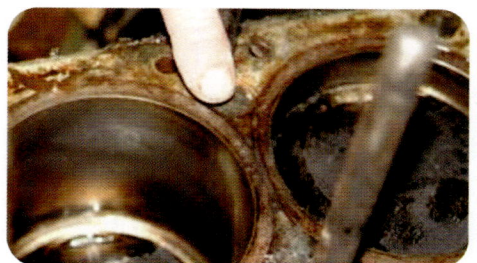

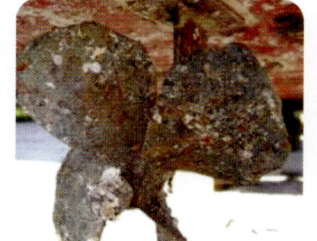

gaskets, due to poor torque and a bad ground. Trouble is when you discover it is way too late and the power head is often wrecked.

A poor lapping procedure between the shaft and propeller can destroy both elements with fatal consequences.

According to with ABYC aluminum fuel tanks never can be installed with copper alloy fittings. Aluminum fittings or Stain Steel series 500 is a good choice.

# Electolytic Corrosion

Electrolytic corrosion is a process of accelerated corrosion. In this process, a metallic surface is continuously corroded by other metals it is in contact with, due to an electrolyte and the flow of an electrical current between the two metals, caused by an external source of current.

The movement of electrons facilitates the division of molecules and the creation of new molecules with different atoms from the seawater.

In simple terms, Electrolytic Corrosion is basically Galvanic Corrosion plus an additional current source. The extra current accelerates the corrosion process between both metals.

This form of corrosion causes widespread damage to critical equipment, and ways and means of monitoring, controlling, and preventing this corrosive damage have been developed and implemented.

Those external sources of current are basically the current leaks that we analyze in the previous chapter. In other words, a poor boat bonding system plus some permanent leaks of current will accelerate the electrolytic corrosion with catastrophic consequences.

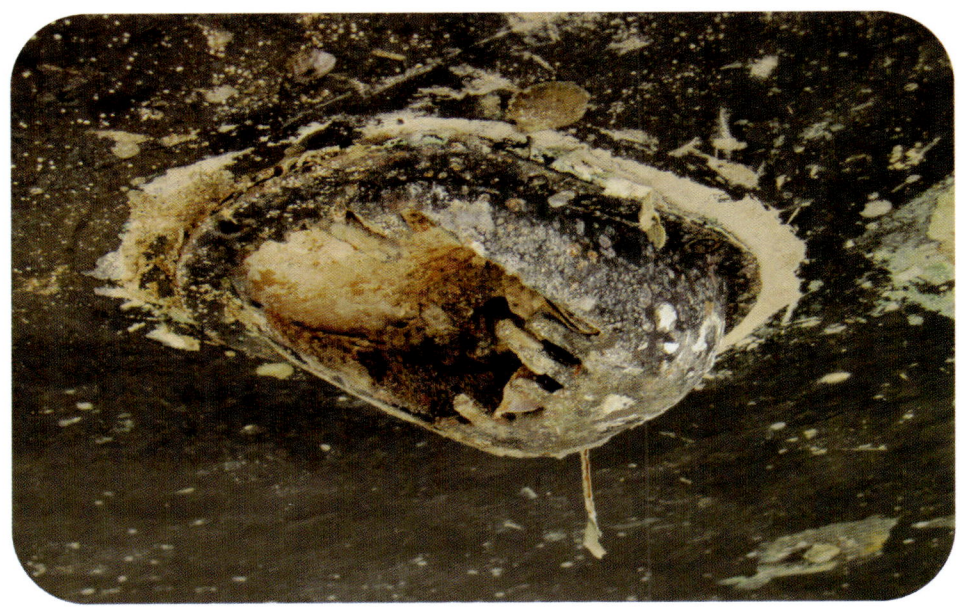

# General Attack Corrosion

Is the most common type of corrosion and is caused by a chemical or electrochemical reaction that results in the deterioration of the entire exposed surface of a metal. Ultimately, the metal deteriorates to the point of failure.

Also referred to as "general corrosion" or "<u>uniform corrosion</u>," general attack corrosion proceeds more or less uniformly over an exposed surface without appreciable localization.

The mechanism of the attack typically is an electrochemical process that takes place at the surface of the material.

Differences in composition or orientation between small areas on the metal surface create anodes and cathodes that facilitate the corrosion process. The surface effect produced by most direct chemical attacks (e.g., as by an acid) is a uniform etching of the metal.

# Localized Corrosion

Localized corrosion is classified as one of three types.

- **Pitting:** Pitting results when a small hole, or cavity, forms in the metal, usually as a result of de-<u>passivation</u> of a small area.
- **Crevice corrosion:** This type of corrosion is often associated with a stagnant micro-environment, like those found under gaskets washers and clamps.
- **Filiform corrosion**: Occurring under painted or plated surfaces when water breaches the coating. filiform corrosion begins at small defects in the coating and spreads to cause structural weakness.

# Pitting Corrosion

It is where the oxide surface breaks down, developing crevices where the corrosion can persist.

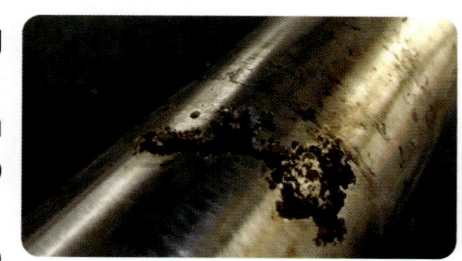

Pitting corrosion or crevice is often encountered in threads and between surfaces where dirt can build up creating a galvanic reaction.

The correct installation of Zinc anode blocks is also important and will significantly reduce pitting corrosion.

# Crevice Corrosion

This form of attack is generally associated with the presence of small volumes of stagnant solution in occluded interstices, beneath deposits and seals, or in crevices, e.g. at nuts and rivet heads.

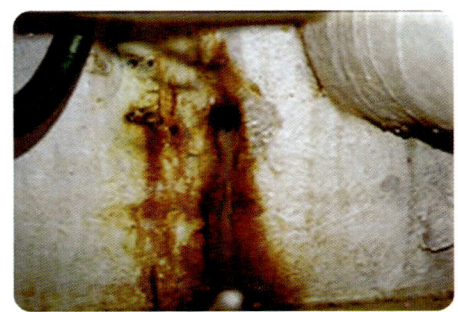

# Filiform Corrosion

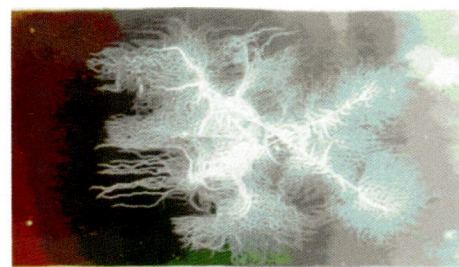

Filiform Corrosion occurs on metallic surfaces coated with a thin organic film that is typically 0.1 mm thick.

Is characterized by the appearance of fine filaments emanating from one or more sources in semi-random directions.

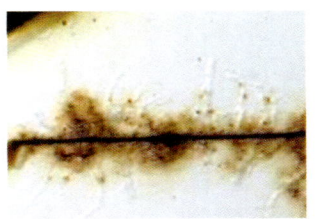

This is a type of corrosion that is commonly known as "localized" and is normally linked to magnesium and aluminum alloys that utilize an organic form of coating.

This type of corrosion has a tendency of taking place in conditions with a high level of humidity. Nitrates, sulfates, carbonates, and condensates that contain halides have been associated with filiform corrosion.

# Environmental Cracking

Is a corrosion process that can result from a combination of environmental conditions affecting the metal. Chemical, temperature, and stress-related conditions can result in the following types of environmental corrosion:

- Stress Corrosion Cracking (SCC)
- Corrosion fatigue
- Hydrogen-induced cracking
- Liquid metal embrittlement

# Stress Corrosion Cracking

Stress corrosion cracking (SCC) is the cracking induced from the combined influence of tensile stress and a corrosive environment. The impact of SCC on material usually falls between dry cracking and the fatigue threshold of that material.

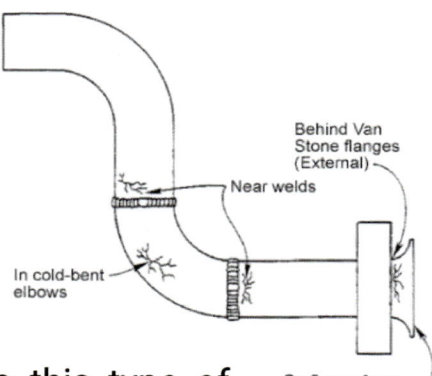

The required tensile stresses may be in the form of directly applied stresses or in the form of residual stresses. The aircraft parts are commonly exposed to this type of corrosion.

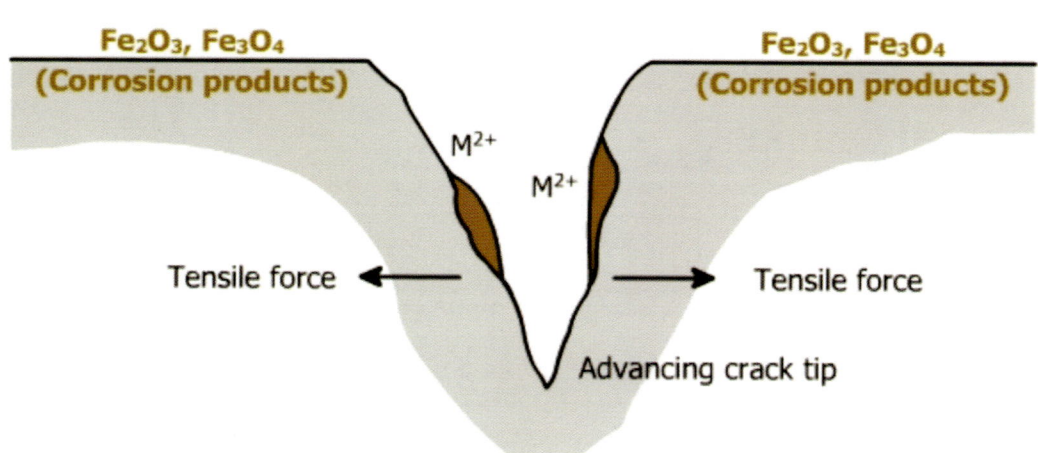

Schematic of stress corrosion cracking.

# Corrosion Fatigue

**Corrosion fatigue** is **fatigue** in a corrosive environment. It is the mechanical degradation of a material under the joint action of **corrosion** and cyclic loading. Nearly all engineering structures experience some form of alternating stress and are exposed to harmful environments during their service life.

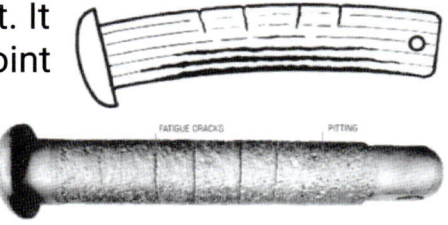

Corrosion fatigue and fretting are both in this class. Much lower failure stresses and much shorter failure times can occur in a corrosive environment compared to the situation where the alternating stress is in a non-corrosive environment.

# Hydrogen-Induced Cracking (HIC)

Hydrogen-induced cracking (HIC) refers to the internal cracks brought about by material trapped in budding hydrogen atoms.
It involves atomic hydrogen, which is the smallest atom, that diffuses into a metallic structure. In the case of a crystal lattice becoming saturated or coming into contact with atomic hydrogen, many alloys and metals may lose their mechanical properties.

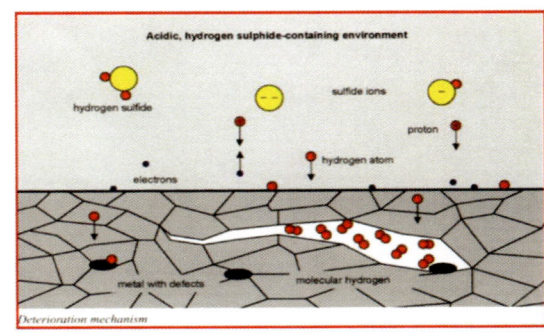

Metallic storage tanks are a good example of HIC, typical failures occurred when storage tank roofs have become saturated with hydrogen by corrosion and then been subjected to a surge in pressure, resulting in the brittle failure of circumferential welds.

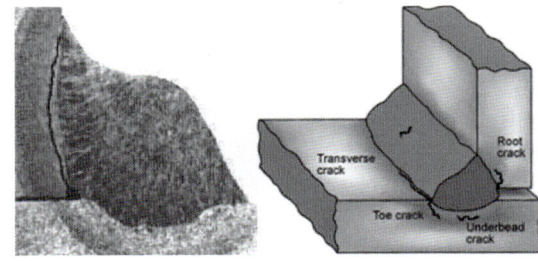

# Flow Assisted Corrosion

Flow-assisted corrosion, or flow-accelerated corrosion, results when a protective layer of oxide on a metal surface is dissolved or removed by wind or water, exposing the underlying metal to further corrode and deteriorate.

- Erosion-assisted corrosion
- Impingement
- Cavitation

# Fretting Corrosion

Occurs as a result of repeated wearing, weight, and/or vibration on an uneven, rough surface. Corrosion, resulting in pits and grooves, occurs on the surface.

Fretting corrosion is often found in rotation and impact machinery, bolted assemblies, and bearings, as well as on surfaces exposed to vibration during transportation.

# High-Temperature Corrosion

Fuels used in gas turbines, diesel engines and other machinery, which contain vanadium or sulfates can, during combustion, form compounds with a low melting point.
These compounds are very corrosive towards metal alloys normally resistant to high temperatures and corrosion, including stainless steel.

High-temperature corrosion can also be caused by high-temperature oxidization, sulfidation, and carbonization.

# CHAPTER
# 8

## Hull Corrosion

## TOPICS

## Video Chapter 8 EP 1: Hull Corrosion Analysis

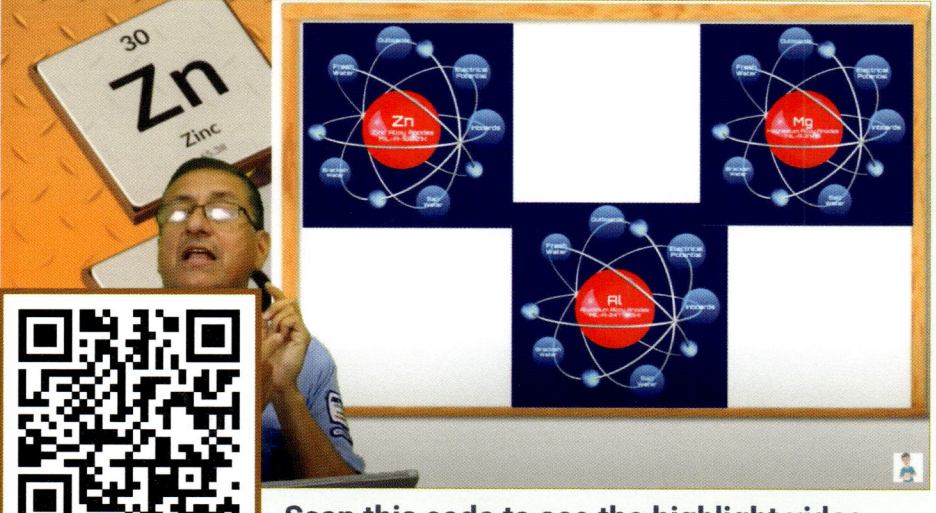

In this video, we will analyze how the hull can be affected by corrosion depending on the hull material, the type of water, the bonding system, and the coating protection. This is a deep analysis to understand how the corrosion process works in different scenarios. Enjoy it !!!

Scan this code to see the highlight video

**Follow me**

# Hull Corrosion

There are many types of corrosion that can affect the hull material. The two basic types are <u>erosion</u> and <u>electrochemical.</u>

## Common Types of Corrosion

Erosion Corrosion = Impingement damage
Crevice Corrosion = Stagnant solution
Poultice Corrosion = Dirt and debris accumulate.
Pitting Corrosion =  localized form of corrosion by which cavities or "holes.
Galvanic Corrosion = Is an electrochemical process in which one metal corrodes preferentially to another.
Intergranular Corrosion= Corrosion in the neighbor of a welding area.

## Erosion on the Hull

Is a strictly mechanical form of corrosion that is caused by friction.
This can be mechanical corrosion, such as that of sandy water flowing around the hull, which acts just like sand paper.

# High Speed Flow Corrosion

There is another type of erosion, which is caused by high-speed water flow.

The pitting one sees on rudder blades behind propellers is an example of non-abrasive erosion.

This is caused by the stream of bubbles from the propeller hitting the rudder. High-speed flow corrosion is rarely found in boats, other than this instance.

# Electrochemical Corrosion

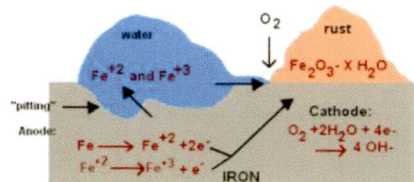

All types of corrosion except mechanical erosion are electro-chemical in nature.

Electrochemical corrosion involves two half-cell reactions; an oxidation reaction at the **anode** and a reduction reaction at the **cathode.**

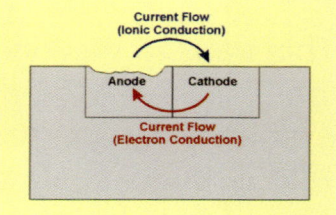

# Equal Electronegativities

If two different materials have the very same electrical charge, nothing will happen. These materials or substances are, we say, "compatible" as in joining certain types of stainless steel and bronze together.

## Electro-Chemical Process

**Electrolysis** is simply the result of stray current, and nothing else.

**Galvanism** is the term applied to the flow of electrons when two dissimilar metals are mated together.

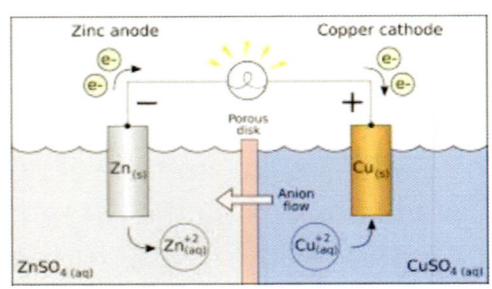

## How the Zinc Anodes Works

The zinc reverses the normal flow of current between dissimilar metals.

The zinc will emit a current that raises and equalizes the electrical potential of all the metals in the system.

It does this by releasing electrons, which are positively charged ions of the metal itself.

This causes the zinc to erode and disappear. These ions will attach themselves to the other metals, which explains why your props and other metals may end up with a rough, scaly surface; they've become covered with zinc oxide.

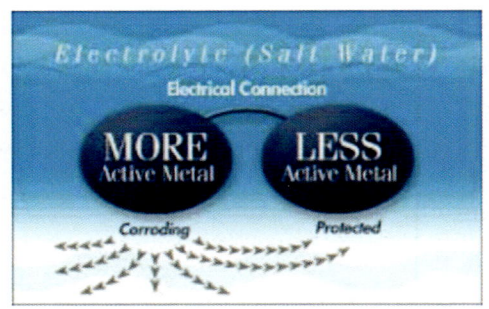

# Crevice Corrosion

As its name implies, crevice corrosion involves water, metals, and crevices.

For our purpose, a crevice is any cavity that will trap and hold water, while at the same time reducing or eliminating air exposure to the water/metal interface.

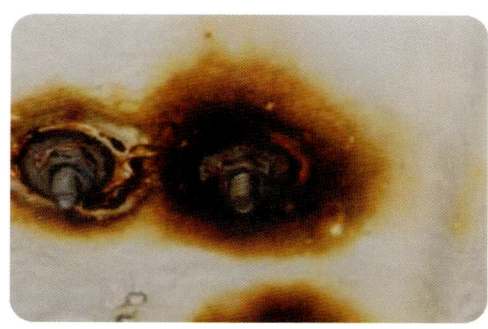

# Crevice Corrosion on Screws

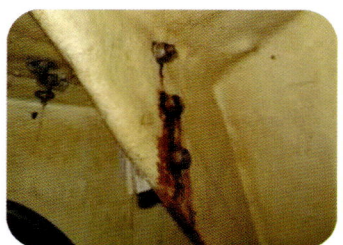

The one on the right shows the typical wasting of the shank right under the screw head. The one on left was exposed to water both under the head and on the inside of the hull where it has thinned at both locations.

The water/metal interface results in oxidation of the metal which concentrates the hydrogen content of water, and turns the water into acid.

This changes the electrical makeup of the affected materials, generating an electrical current that "dissolves" the metal involved.

# Fiberglass Hull Corroded

These crevices or closed cells can become dynamic, meaning that the process can perpetuate itself for a long time -- either until the acidic water is exhausted.
The corrosion process continues until the metal is completely gone.

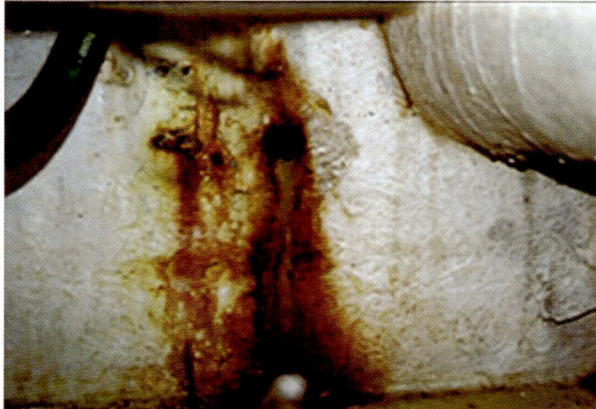

## Stress Corrosion

This is a combination of crevice corrosion cells combined with heavy loading.
It most often occurs on sailboat rigging and powerboat propeller shafts and propellers.

## Stress Corrosion on Shafts

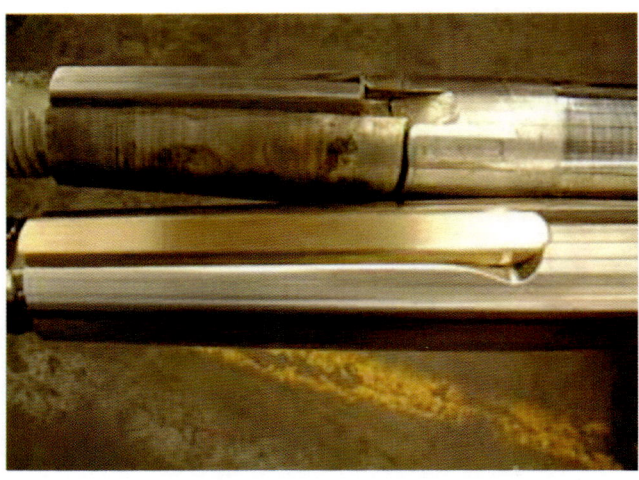
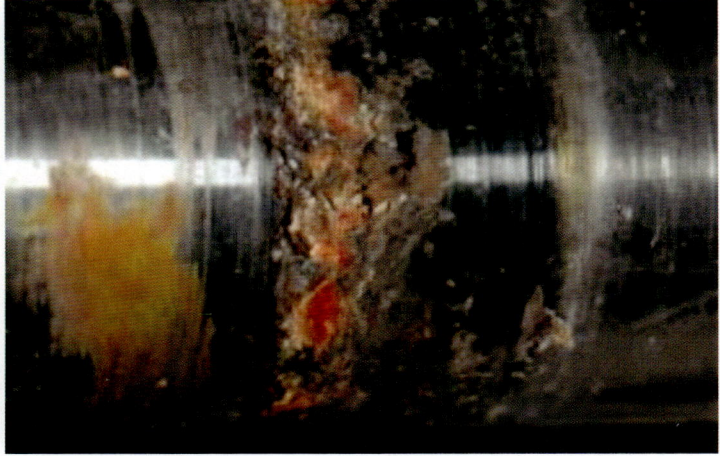

## Hardware on Sail Boats

Old-style swage fittings on sailboat rigging combine both stress and corrosion cells from entrapped water within the swaged cable.

# Stainless Steel Fasteners

## Video Chapter 8 EP 2: Sacrificial Anodes Selection Criteria

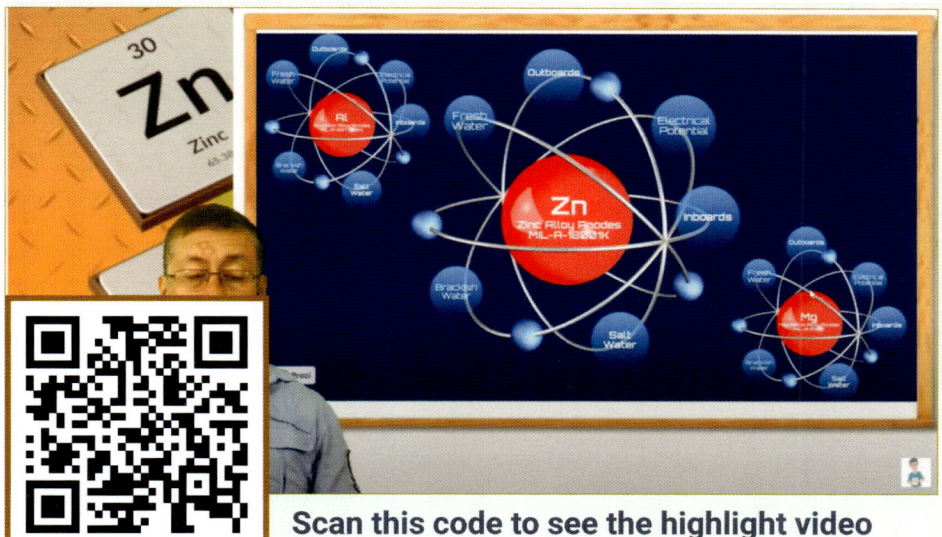

Scan this code to see the highlight video

In this video, you will find the procedure to select the appropriate sacrificial anode for your boat according to the hull material, the type of water, and the propulsion system in your boat. This is a great video for boat owners and boat enthusiasts.
Enjoy it

Follow me

## Stainless Steel Fasteners

18/8 stainless steel, also known as type 302 or 304-grade stainless steel, is nominally 18% chromium and 8% nickel, with the remainder being mainly iron.

Type 302 has the same corrosion resistance as 304, with slightly higher strength due to additional carbon.

Type 304 is the most common grade; the classic 18/8 stainless steel. Outside of the US, it is commonly known as "A2" Stainless. (see our Stainless Steel Metric Fasteners).

Type 316 is the second most common grade (after 304); for food and surgical uses.

## Metallic Hull's

All metal boats are potential candidates for galvanic corrosion.

Protective coatings can help prevent the build-up of organic material on your hull.

Organic material can be an aid to corrosion.

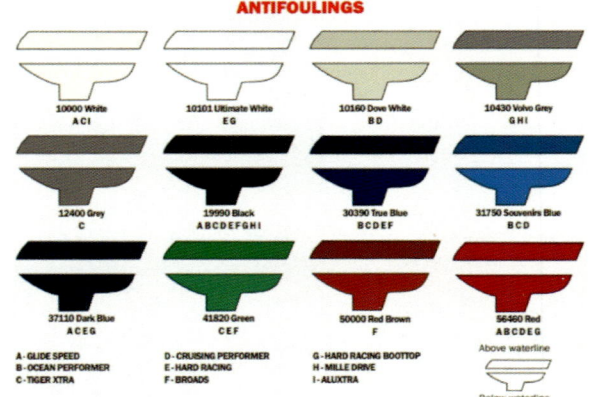

## Aluminum Hull's

If you use a protective bottom paint make sure it is made for use with aluminum. It must not be based on a dissimilar metal. copper is particularly problematic.

Copper pennies, copper shell casings, the heavy copper wire used as fish clubs have all been known to be starting points for corrosion if left in the bottom of an aluminum boat.

Watch for corrosion near bolt holes, chipped paint, behind peeling paint, and any place the aluminum has been scratched or dented.

Inspect the keel, chine extrusions, and corner castings regularly. If they have been dented or scratched on rocks or beaches they may provide a starting point for electrolysis.

# Aluminum Hull's

Except for sacrificial anodes, don't mount any metal to the hull other than aluminum or high-quality stainless steel.

If you use fasteners they should be plastic or stainless, not zinc, brass, copper, or iron-based. Inside the boat, don't let any metal sit in the same spot in the bilge for long.

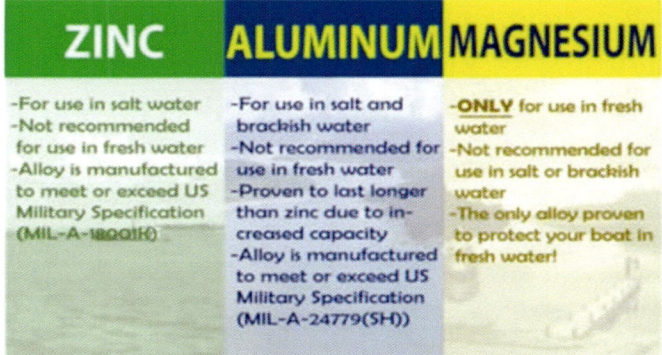

| ZINC | ALUMINUM | MAGNESIUM |
|---|---|---|
| -For use in salt water -Not recommended for use in fresh water -Alloy is manufactured to meet or exceed US Military Specification (MIL-A-18001H) | -For use in salt and brackish water -Not recommended for use in fresh water -Proven to last longer than zinc due to increased capacity -Alloy is manufactured to meet or exceed US Military Specification (MIL-A-24779(SH)) | -**ONLY** for use in fresh water -Not recommended for use in salt or brackish water -The only alloy proven to protect your boat in fresh water! |

# Aluminum Hull's (Bonding)

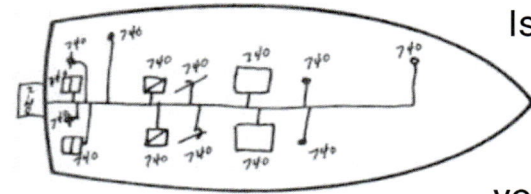

Isolate the hull from electrical current. Make sure your battery is grounded to your motor. Don't ground electrical devices to the aluminum hull.

Don't let salt residue build upon any surfaces. If you see corrosion arrest it immediately before small damage becomes big damage.

# Aluminum Hull's (Cathodic Protection)

Cathodic protection is supplied by the use of sacrificial anodes. Fortunately for aluminum, there are metals that are even less noble. They include magnesium, zinc some aluminum, zinc alloys.

The anodes when attached directly to an unpainted area of an aluminum hull will protect any part of the hull that is not electrically isolated from the anode.

**Anodes**

| | |
|---|---|
| -1600 | Magnesium |
| -1130 | Zinc |
| -900 | Chrome |
| -850 | Aluminium |
| -700 | Cadmium |
| -610 | Steel |
| -510 | Lead |
| -500 | Stainless steel (active) |
| -360 | Copper, Bronze |
| -310 | Tin |
| -150 | Titanium |
| -80/-50 | Stainless steel (Passive) |

**Cathodes**

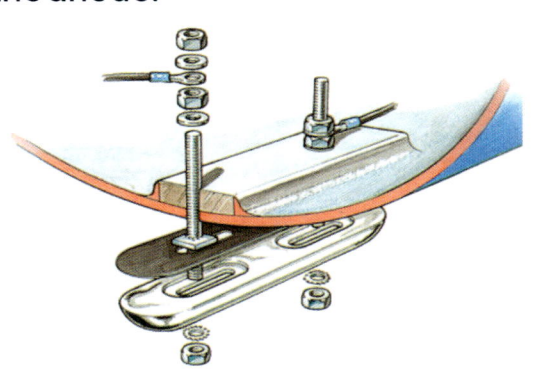

## Aluminum Hull's (Cathodic Protection)

Which one is best?  for our boats probably aluminum. zinc allow.   Magnesium is very active in saltwater and is sometimes used for fresh water applications.

Zinc works well but can oxidize over when exposed to air.

| | Zinc | Navalloy Aluminum | Magnesium |
|---|---|---|---|
| **Voltage** (in seawater) | -1.03V | -1.1V | -1.6V |
| **Relative Life** (Zinc = 100 Same Size) | 100 | 150 | 30 |
| **Relative Density** (Zinc = 100) | 100 | 42 | 27 |
| **Mil.Spec.** | MIL-A-18001 | MIL-A-24779 | MIL-A-21412 |

This means if zincs are not cleaned when a boat has been out of the salt and the boat is returned to the salt the oxidized coating will protect the zinc and it will not function as a sacrificial anode.

Some aluminum, zinc alloys will function well and not oxidize over so they are better for boats that may be periodically out of the saltwater.

Anodes do not last forever.  Somewhere after 1/3 to 1/2 of the metal in an anode has disappeared it is time to replace it.  Some anodes will have built-in warning devices that indicate when replacement is necessary.

## Question

**My outboard has an anode on it. Will that protect my boat?**

- If the zinc is sufficient in size, the motor is connected electrically to the hull and the zinc is always in the water it may protect the hull.  When the motor is lifted out, the anode is no longer in the circuit with a hull that remains in the water and there is no protection.  The hull should have its own anode to be safe.

**Is my boat safe from corrosion when it is on the beach?**

- If it is embedded in the sand that is often wet there may be some risk.  If saline water collects inside the boat there may be some risk.  An anode mounted outside the boat will not protect metal inside the boat if the anode is not submerged in the water that has collected inside.  A separate anode might be desirable for inside the boat.

## Coating Systems & Corrosion protection

## TOPICS

### Video Chapter 9 EP 1: Marine Naval Alloys & Surface Coatings Processes

In this video, we will analyze the different types of Naval Alloys used in marine applications as well as the surface coatings process recommended for our industry. I hope this learning material be useful in your business.

**Scan this code to see the highlight video**

**Follow me**

# Coating Systems

Coating systems can be divided into two broad groups: **Metallic** and **Non-metallic.**
The Metallic coatings can be separated into two groups:
<u>Barrier coatings</u>.- Protect the substrate metal by preventing moisture from penetrating the surface.
<u>Sacrificial Coatings</u>.- Provide some degree of barrier protection but they combine that with cathodic protection.

# Thermal Barrier Coatings

These coatings serve to insulate components from large and prolonged heat loads by utilizing thermally insulating materials which can sustain an appreciable temperature difference between the load-bearing alloys and the coating surface.

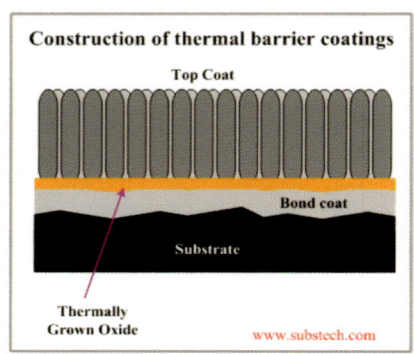

# Hot-Dip Coatings

Hot –Dip Coating is used to apply coatings of Zinc, Aluminum, Lead, Tin, and some alloys of these metals.
<u>Galvanizing</u> generally refers to <u>hot-dip galvanizing</u> which is a way of coating steel with a layer of metallic zinc.

It is the process of coating <u>iron</u>, <u>steel</u>, or <u>aluminum</u> with a layer of <u>zinc</u> by immersing the <u>metal</u> in a bath of molten zinc at a temperature of around 860 °F (460 °C).
**Galvanized steel** is widely used in applications where <u>corrosion</u> resistance is needed without the cost of <u>stainless steel.</u>

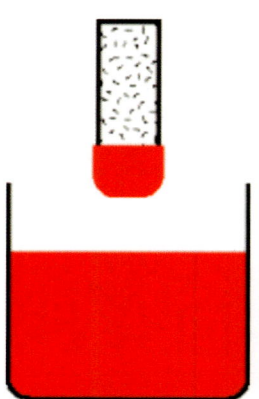

# Zinc Metal Coating (Cold)

Zinc works as a barrier metal coating and as a sacrificial metal coating to prevent oxidation.
If the zinc coating is scratched or penetrated, it continues to provide protection by galvanic action until the zinc layer is depleted.

# Aluminum Metal Coating

**Aluminized steel** is steel that has been hot-dip coated on both sides with aluminum-silicon alloy. This process assures a tight metallurgical bond between the steel sheet and its aluminum coating, producing a material with a unique combination of properties possessed neither by steel nor by aluminum alone.

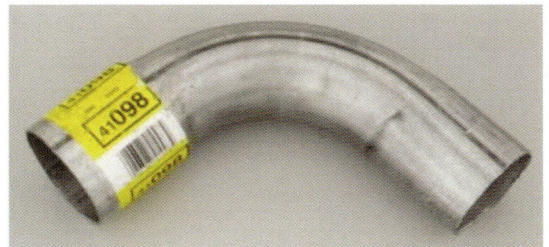

It is commonly used for heat exchangers in residential furnaces, commercial rooftop HVAC units, automotive mufflers, ovens, kitchen ranges, water heaters, fireplaces, barbecue burners, and baking pans. This steel is very useful for heating things up because it transfers heat faster than most other steel.

# Sherardizing

Sherardizing involves heating the articles to be coated in zinc powder to approximately 400°C at which temperature diffusion bonding of the zinc with the steel occurs.

The coating thickness is limited to about 1.5 mils (0.038 mm).

# Mechanical Coating

Mechanical plating involves tumbling the items to be coated in zinc powder with glass beads and special reducing agents to bond the zinc particles to the steel surface.

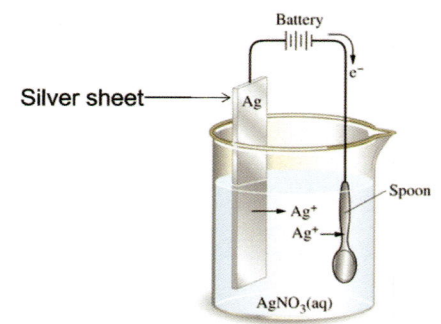

As the barrel rotates, zinc dust is added from an overhead feed hopper. Peened by the impact from the glass beads, the fine zinc dust is cold-welded to the tumbling parts

By controlling the amount of zinc, the plating thickness can be varied from 0.1 to 5 mils ( 0.0025 to 0.127 mm).

# Zinc Electroplating

Also known as electrodeposition, electroplating is using a small sheet of metal in an electrolytic cell to coat another object. It is used to protect objects from damage against rusting and corrosion of metals.

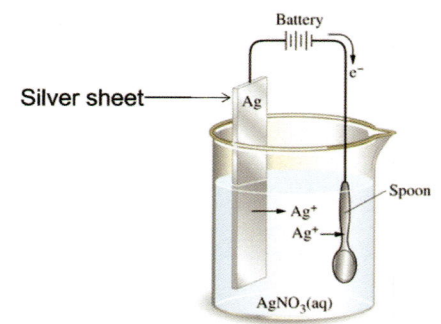

# Common Uses

Electroplating is commonly used for plating jewelry that uses silver or gold, silverware which is most commonly uses silver, and vehicle parts that use chromium.

# Benefits of Electroplating

Tin is used to coat cans in order to prevent corrosion, while a chromium coating will increase a metal's resistance to wear.
Silver and Gold are used to coat silverware and jewelry to prevent corrosion and increase value.

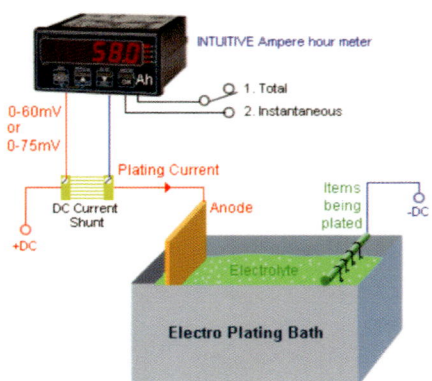

# Zinc Metal Spraying

Zinc met metal spraying requires the steel surface to be cleaned to a Class 3 level and then zinc wire or zinc powder is sprayed onto the surface with an oxy-acetylene or plasma flame gun.

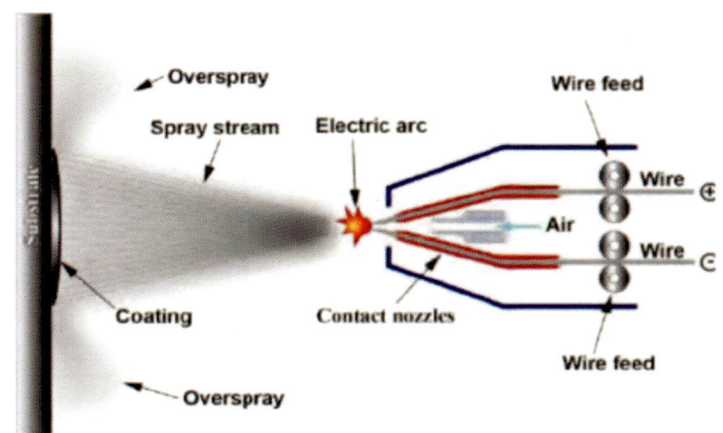

Thermal spraying usually involved several stages:

- Cleaning of the surface from oil and other contaminants.
- Surface preparation usually consists of abrasive grit blasting to roughen the surface.
- Thermal spraying: The sprayed metal impacts and attaches to the roughened steel and solidifies. Subsequent layers of the spray should build up the coating thickness to as thick as 300 microns.
- Sealing of the coating unless it is to be painted to provide additional protection to the steel.

# Twin Wire Arc Spraying

Two wires (hence a common term for the process is Twin Wire Arc Spray), are fed into the pistol and electrically charged, one positive and one negative.
The wires are forced together and form an electric arc, melting the wire. Compressed air, passing through a nozzle, atomizes the molten metal and sprays it onto the work piece.

# Flame Spraying

The gas fuel and oxygen are mixed and ignited to produce a flame. The material, either a wire or powder is fed into the flame.

For wire flame spray, the material is melted and the compressed air, passing through a spray nozzle atomizes the molten metal and sprays it onto the workpiece. The larger the wire diameter, the higher the spray rate.

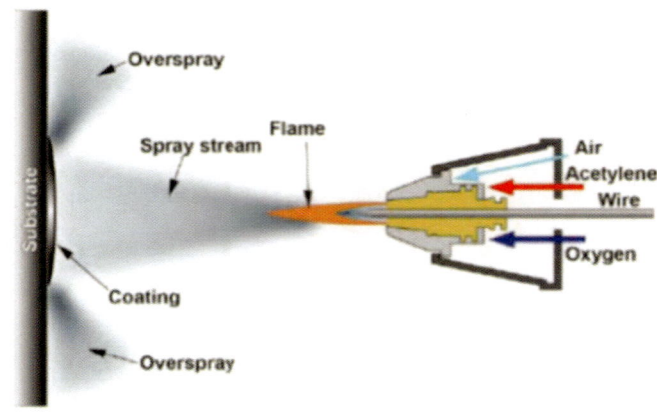

# Surface Preparation

To guarantee good bond strength between the steel target and the sprayed metal coating, a specified roughness must be applied to the steel surface.

The purpose of the preparation is to remove surface imperfections such as oxides of iron and to provide a surface profile on the steel target that has angular peaks and valleys, caused by the impact of the abrasive.

# Sealing The surface

Sealing is very different from painting. Sealers are of low viscosity and penetrate into the pores of the sprayed coating.

The sealer gives a smoother surface, reducing the pickup of dirt and other contaminants, preserving the surface appearance, and reducing maintenance costs.

**Video Chapter 9 EP 2: The Impressed Current Cathodic Protection Process**

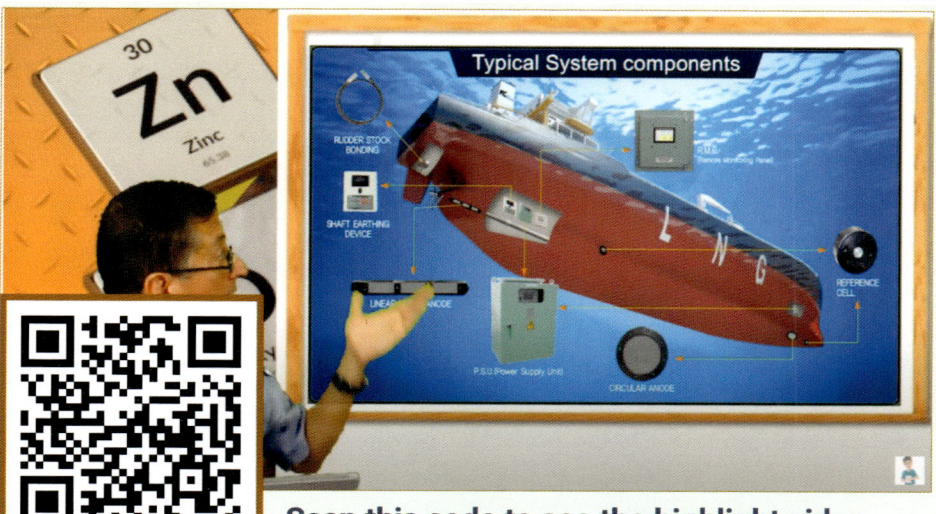

In this Video you will learn who the ICCP system works and why this system is more popular in the marine industry each day. Enjoy it

Scan this code to see the highlight video

Follow me

# Cathodic Protection Systems

All cathodic protection systems require an anode, a cathode, an electric circuit between the anode and cathode, and an electrolyte.
cathodic protection will not work on structures exposed to air environments. The air is a poor electrolyte, and it prevents current from flowing from the anode to the cathode.

# Types of Cathodic protection

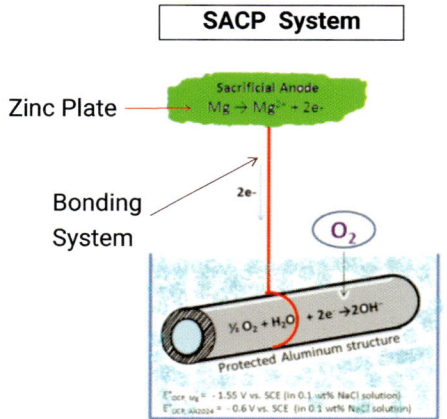

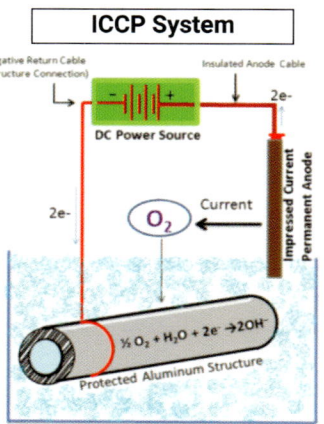

There are two methods for cathodic protection:

1. Impressed current cathodic protection (ICCP)
2. Sacrificial (galvanic) anode cathodic protection.

# Types of Cathodic protection

The main difference is that **ICCP** uses an external power source with inert anodes where as **SACP** uses naturally occurring electrochemical potential differences between metallic elements.

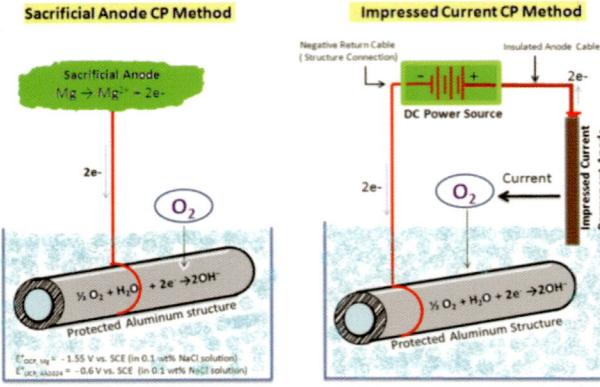

# Sacrificial (galvanic) anode cathodic protection (SACP)

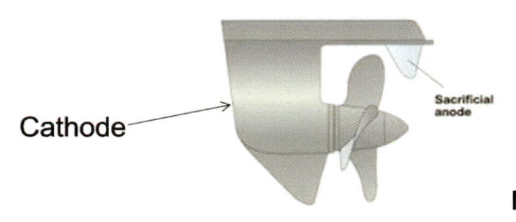

This is the typical situation with an aluminum or Iron structure protected with a more active metal such as zinc or magnesium.

This produces a galvanic cell in which the active metal works as an anode and provides a flux of electrons to the structure, which then becomes the cathode.

The cathode is protected and the anode progressively gets destroyed, and is hence, called a sacrificial anode.

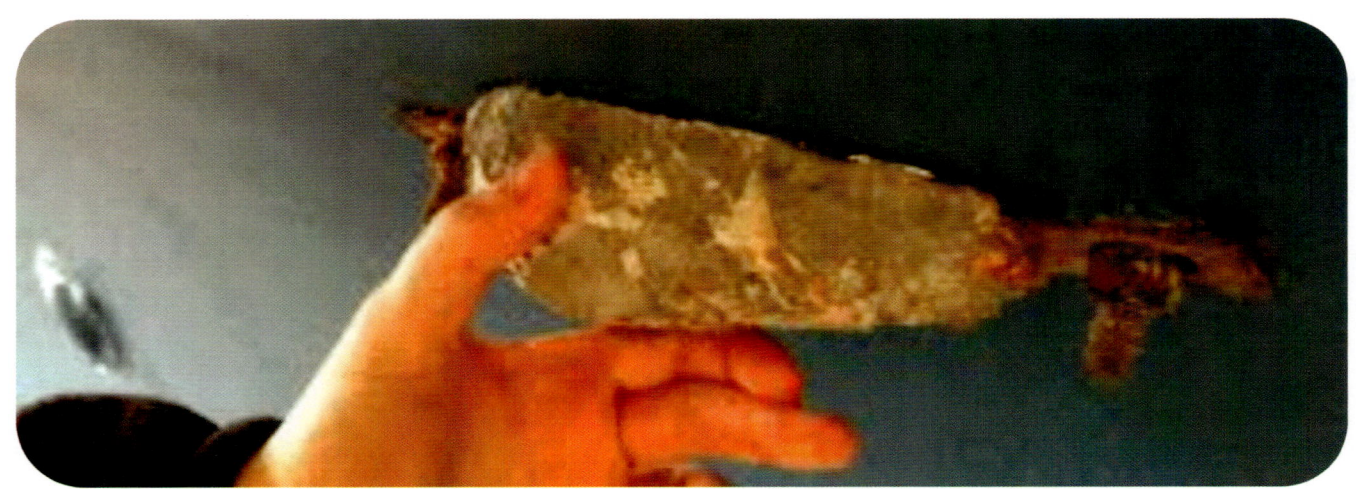

# Impressed Current Cathodic Protection (ICCP)

Impressed current cathodic protection is an external DC current-controlled, applied to the structure of the metal to be protected by means of the negative terminal of the source, while the positive pole is coupled to an auxiliary anode.

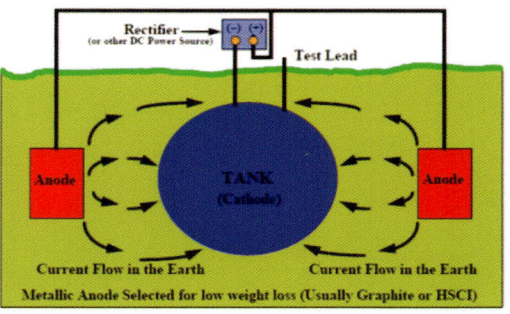

Impressed current cathodic protection works by delivering controlled amounts of DC current to the surfaces submerged in water with the aid of ultra-reliable zinc anodes.

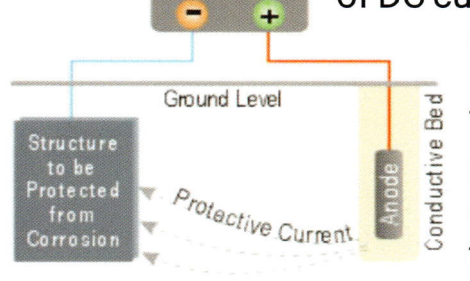

The electrical current that is continuously regulated and monitored by the ICCP system helps prevent the electrochemical mechanism of galvanic corrosion prior to its attack.

## How ICCP Works ?

The rain of electrons from the anode to the structure tries to saturate the structure metal in order to complete 8 electrons in the outer orbit of each atom.
In this way, the atoms on the metallic structure try to be stable and do not react with oxygen.

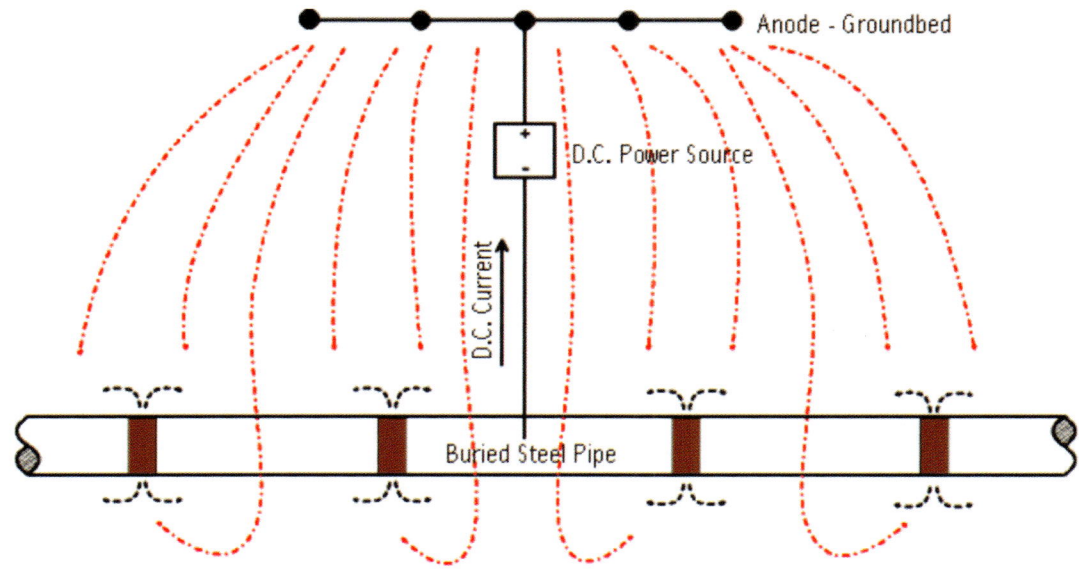

## The Anode

Platinum is an excellent anode material due to its high conductivity and low consumption rate. Due to its high cost, it is not economical to use platinum by itself.

The typical expectation of impressed current anode life is over fifteen years.

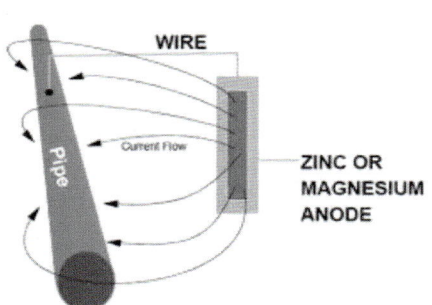

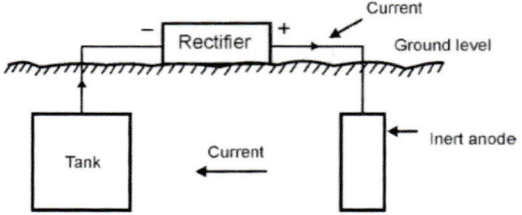

Anode materials that have proven to be suitable for impressed current systems include treated graphite, high silicon cast iron, mixed metal oxide, and platinum and magnetite.

Platinized anodes are produced by electroplating a thin layer of platinum over a lower-cost substrate such as titanium, tantalum, niobium zirconium, etc.

## Impressed Current Cathodic Protection (ICCP)

Since the driving voltage is provided by the DC source there is no need for the anode to be more active than the structure to be protected.

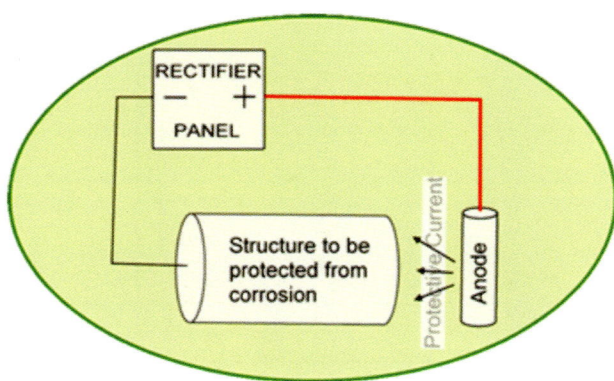

## Uses

This method is widely used in static ocean structures such as pipelines, fuel containers, or big ship cargo containers.

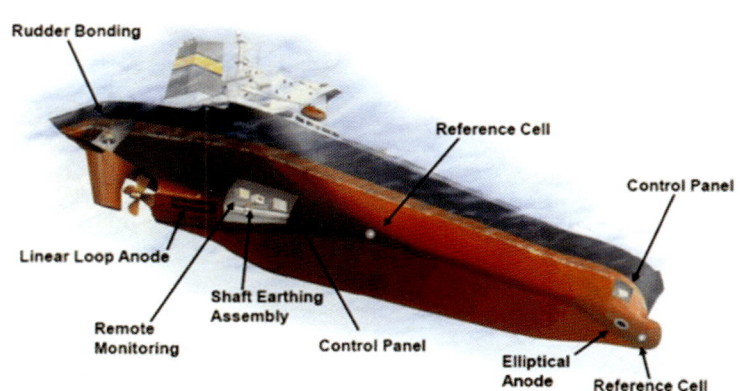

## Preventing Hull Corrosion with ICCP Systems

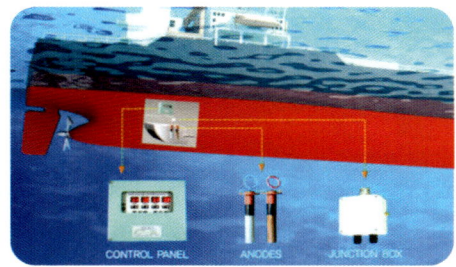

Using an arrangement of hull-mounted anodes and reference cells connected to a control panel, the ICCP system produces a powerful impressed external current to suppress the natural electrochemical activity on the wetted surface of the hull.

## Hull Protection with ICCP

The current to the anodes is constantly monitored and adjusted to provide the "optimum" level of protection at all times. This is far superior to the performance of sacrificial anodes.

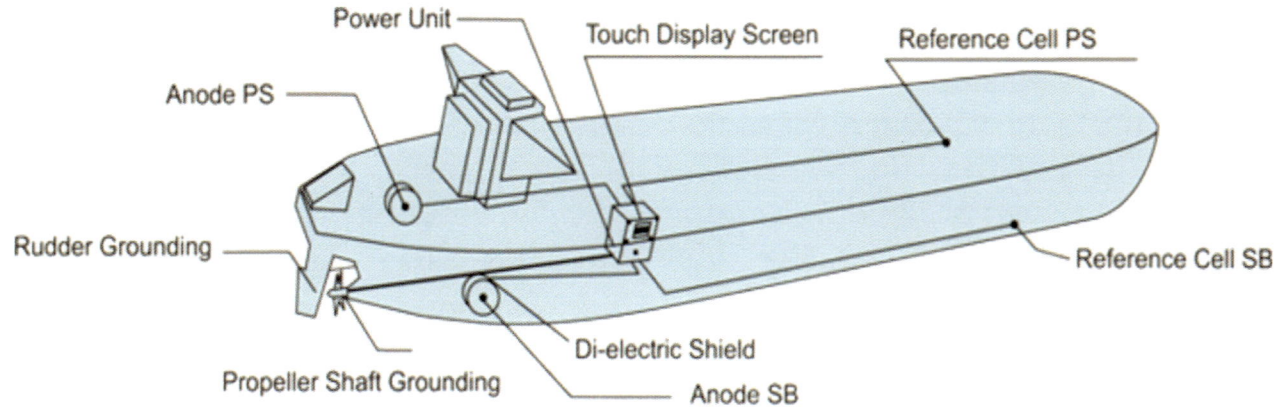

## Protecting Shafts and Propellers

For parts in movement the current entry into the structure by means of contact brushes.

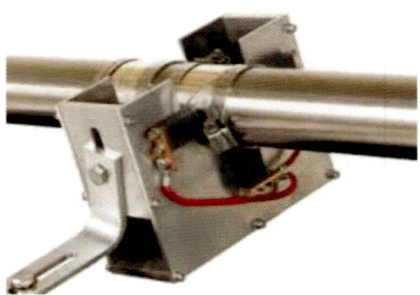

## Advantages

This eliminates the formation of aggressive corrosion cells on the surface of plates and avoids the problems, which can exist where dissimilar metals are introduced through welding or brought into proximity by other components such as propellers.

**More Advantages**

- Anode life 15 years - unlike sacrificial anodes.
- Reduced weight in comparison with sacrificial anodes.
- Flush-mounted anodes to ensure a smooth hull profile.
- No maintenance is required.
- Self-diagnostic system.
- Operates from 230V or 115V AC. electrical supply.
- The Control panel only measures 500 x 390 x 210 mm for systems up to 40 amps.
- Lower yard installation costs than recessed sacrificial anodes.

## MerCathode Protection System

The Mercury Precision MerCathode system provides automatic protection against galvanic corrosion. It is a solid-state device that operates off a boat's 12-volt battery and provides protection by impressing a reverse blocking current that stops the destructive flow of galvanic currents.

# MerCathode Protection System

Similarly, in the case of electrolytic corrosion of pipes, high-voltage direct current equipment, and marine equipment, some external leakage current is always the driving force for the corrosion.

Prevention of this form of corrosion involves devices that break the continuity of the circuit or provide an alternate low-resistance path to connect the leakage current directly to the ground. For example, in the case of motor or generator bearings, a current collector mechanism with a brush holder and a brush makes contact with the rotating shaft, connecting the shaft to the ground. Alternatively, the bearing pedestals may be insulated from the ground to break the path of leakage current.

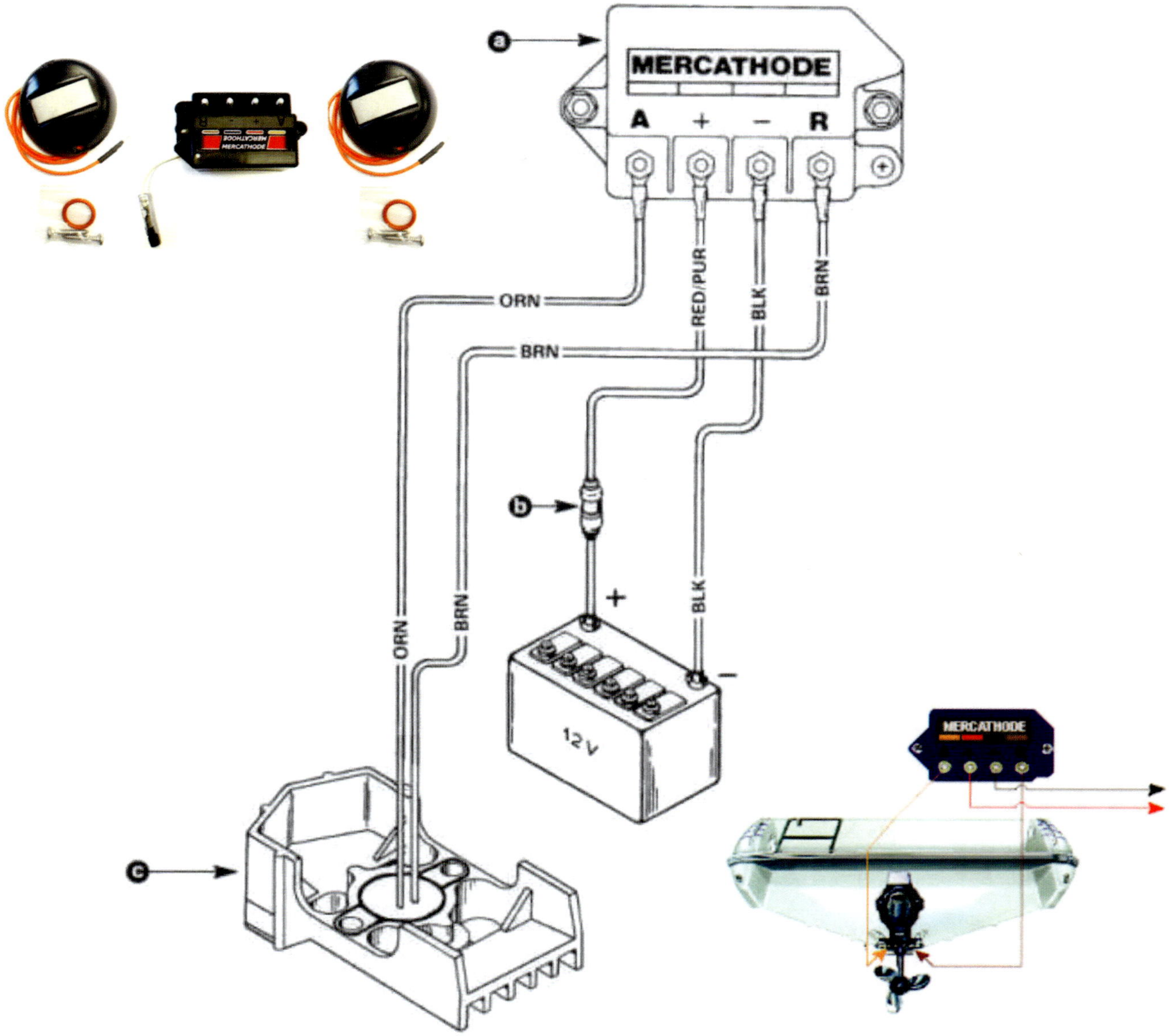

# Metallurgical Processes

## TOPICS

## Sand Casting

Sand casting, also known as a sand molded casting, is a metal casting process characterized by using sand as the mold material.

Most castings, especially large aluminum products, are usually made in sand molds. Casting is the process where metal is heated until molten. While in the molten or liquid state it is poured into a mold or vessel to create a desired shape.

Casting parts are more susceptible to being corroded due to their porous structure in comparison with forging parts.

# Metallurgical Structure Cast Aluminum

The white holes correspond to air bubbles trapped during the foundry.
The black marks correspond to carbides coming from the crucible degradation.

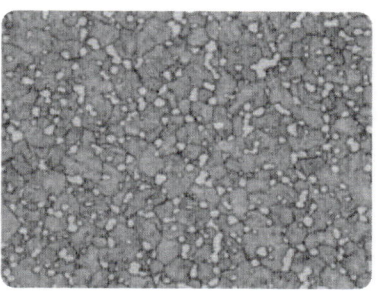

# Stress Corrosion Cracking

Due to the high amount of impurities and high concentration of air bubbles, the molten aluminum is not recommended for parts subjected to cyclic loading because the bubbles are propagated as cracks quickly.

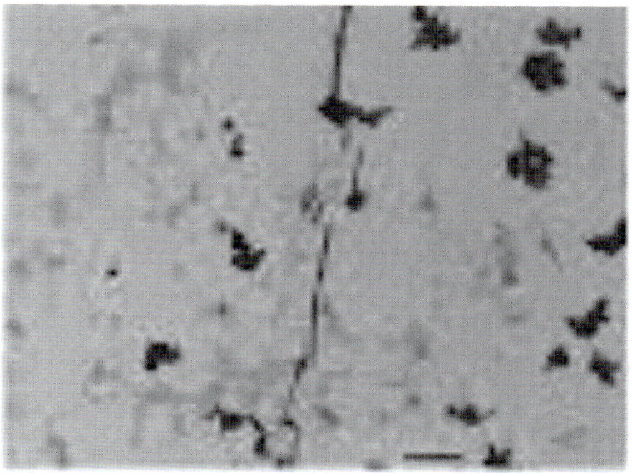

(a)

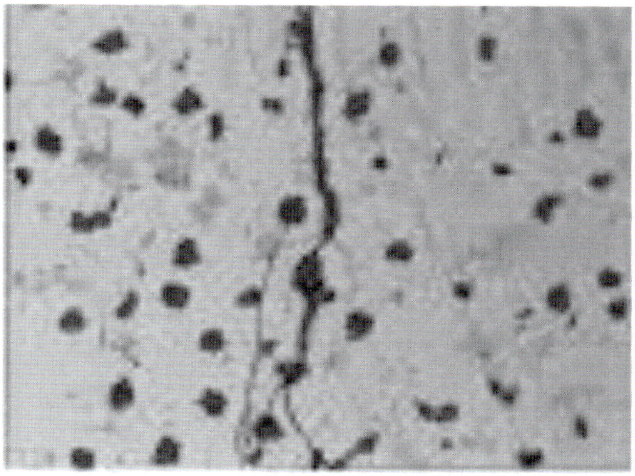

(b)

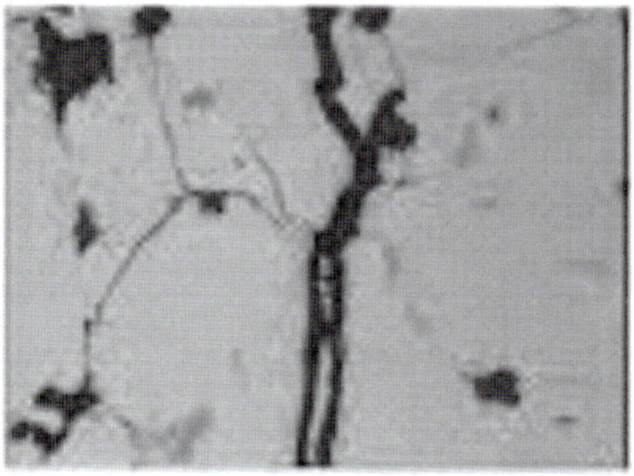

(c)

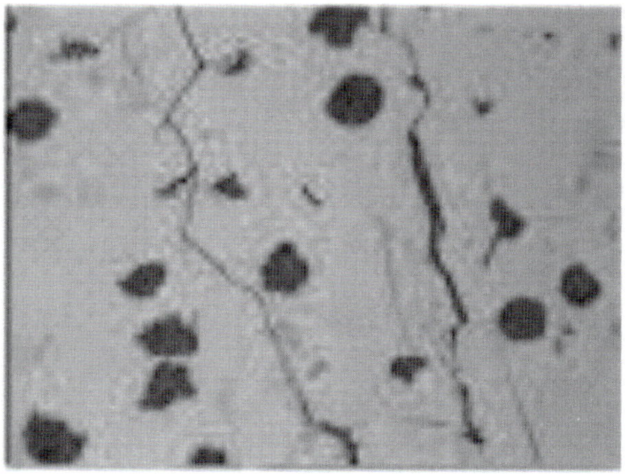

(d)

# Sand Casting

Casting is the process where metal is heated until molten. While in the molten or liquid state it is poured into a mold or vessel to create the desired shape.

Casting is used to make metal components of all sizes, ranging from a few ounces to several tons. Sand molds can be formed to create castings with fine exterior detail, inner cores, and other shapes. Nearly any metal alloy can be sand cast. Hollows are made in moistened sand, filled with molten metal, and left to cool.

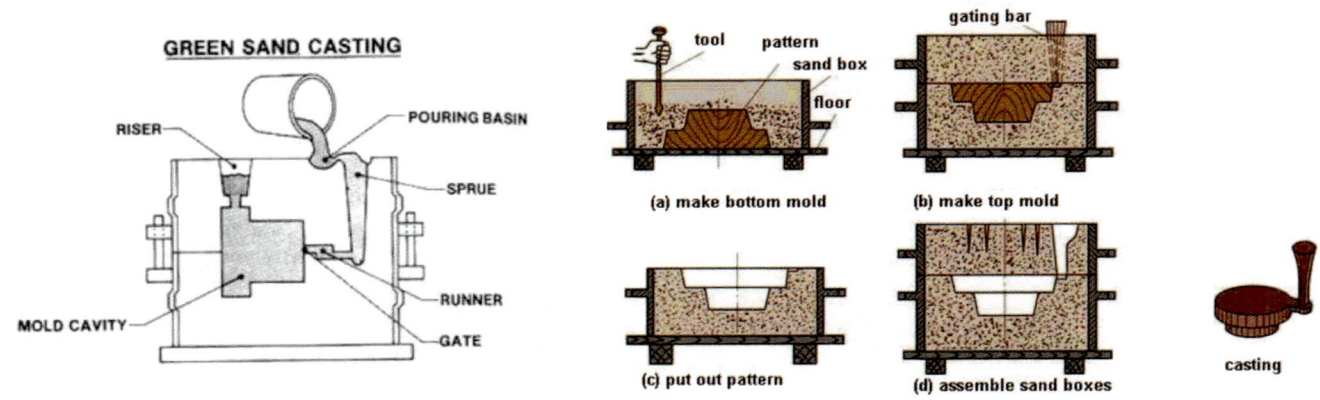

Castings are produced by pouring liquid metal into a mold cavity. For a casting to be successful, the mold cavity must retain its shape until the metal has cooled and fully solidified. Pure sand breaks apart easily, but molding sand contains bonding material that increases its ability to resist heat and hold shape

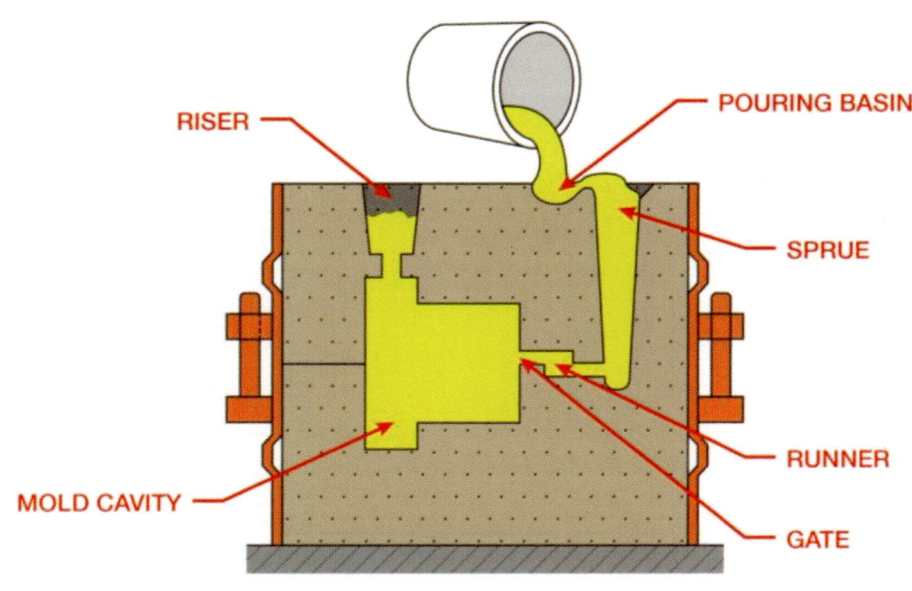

GREEN SAND CASTING

# The Cast Molding Process

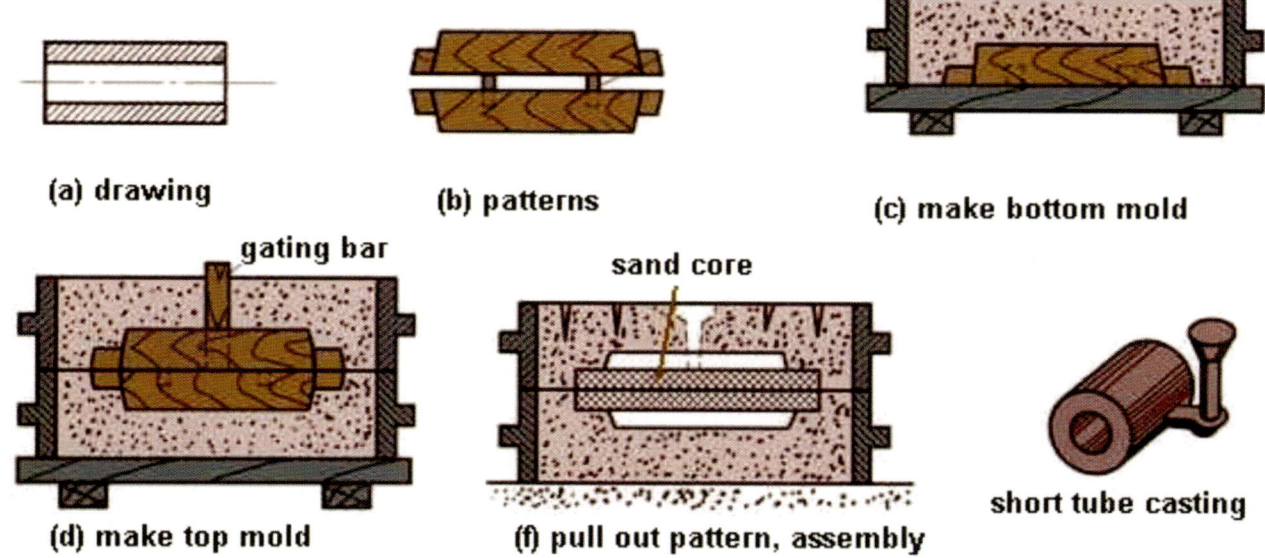

(a) drawing      (b) patterns      (c) make bottom mold

(d) make top mold      (f) pull out pattern, assembly      short tube casting

# The Molding Process

In Sand Casting, a pattern of the desired finished part including the metal delivery system (gates and risers) is constructed out of hardwood, urethane, metal, or foam.

The pattern is removed from the bonded sand, leaving a cavity in the mold that is in the shape of the part. Molten metal is poured into the cavity and the metal solidifies.

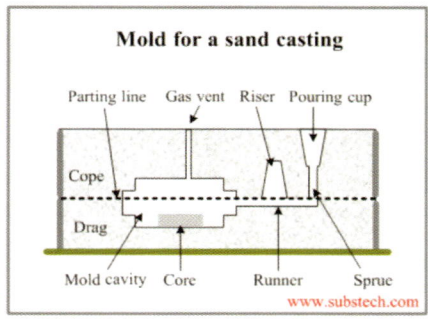

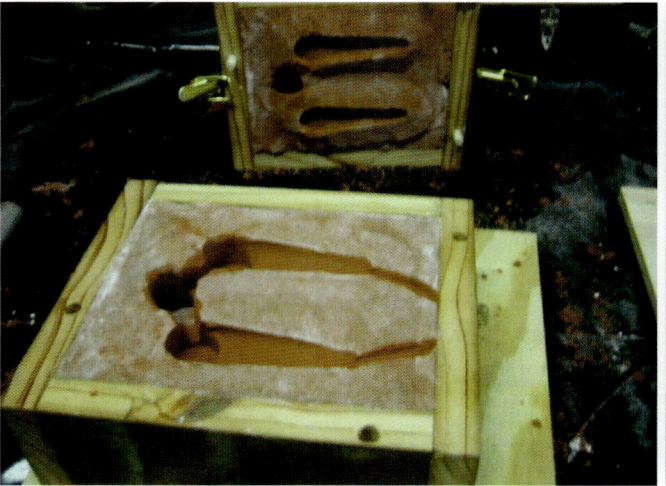

# Molding Sand Properties

Molding is the process of making a cavity or mold out of the sand by means of a pattern. The molten metal is poured into the molds to produce the casting.

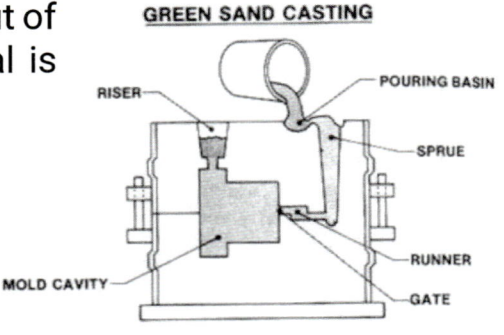

# Foundry Sand/Greensand

Foundry sand consists primarily of clean, uniformly sized, high-quality silica sand or lake sand that is bonded to form molds for ferrous (iron and steel) and nonferrous (copper, aluminum, brass) metal castings.

The sand used in greensand is silica, common/ordinary sand. You can use beach sand, desert sand.

# Properties of Molding sand

### Porosity or Permeability
• It is the property of sand that permits the steam and other gases to pass through the sand mold

### Plasticity
• It is that property of sand due to which it flows to all portions of the molding box or flask. The sand must have sufficient plasticity to produce a good mold

### Adhesiveness
• It is the properties of sand due to it adheres or cling to the sides of the molding box

# The Crucible

There are crucibles of Cast iron, Graphite with Molybdenum, and Ceramic crucible.

# Furnaces

There are two types: LPG and Charcoal coke. Both of them are mixed with air at high pressure from a blower.

A furnace for non-ferrous alloys reaches temperatures around 2800 F.
The melting point of aluminum is 1220 F.

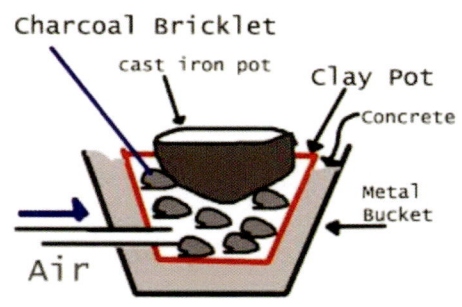

## Sand Casting Tools

Riddle (Sieve/screen), large hole cutter (1/2" copper pipe), dowels (1/4" and 1/8"), rammer/striker (wood 10"x1.5"x1"), runner bar pattern, spoon, trowel and parting dust (in the black sock).

# The Pattern

Place the original metal pattern in the flask (wood box) with enough room for gating. In this mold, we use a wood pattern for "runner" and "gate".

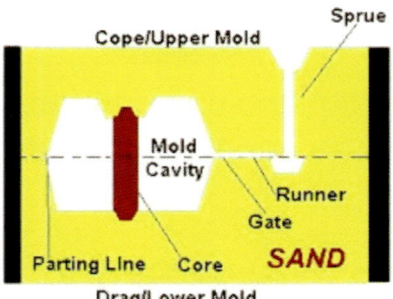

## Dust Pattern

Dust pattern with parting dust to keep it from sticking. Parting dust is a hydrophobic material, it repels moisture. Most powders (baby, talcum) absorb moisture.

## Covering the Mold

Use sifted sand just to cover the mold, then fill up the flask with sand, level (flush) with the top.

## Filling and Packing Box

Use paddle side of rammer to tuck edges first. Hold the flask with your other hand.

Use the butt side of rammer east to west (lightly to protect pattern), then north and south (harder to pack mold tight). Ramming strokes should be 1 inch apart like you are planting corn.

You can ram too soft, you can't ram too hard. Fill it up with sand to about 2 inches above the box.

## Removing Excess Sand

Remove excess sand with a straight metal edge.

# Flipping it Over

Place bottom board on top of your mold. Holding bottom board and flask together, flip it over.

# Preparing for the Upper Box

Remove the cope and pattern board.
Use your spoon and smooth the edges of the pattern and any rough areas.

# The Sprue and Pop-Up (Vent)

Put in place two metal pipes surrounded by sand in order to produce the sprue and pop-up.

## Making the Sprue and Vent

Take cope off and set it aside.
Carve pouring cup into the sprue hole on the top of the mold. Like a small cup. Slick what you can with the spoon.

## Removing the Runner and Gate

Tap runner pattern to loosen, then remove.
Remove wood gate pattern.

## Removing the Pattern

Tap pattern lightly to loosen.
Put a screw partially into the pattern for a handle. Remove pattern.
Smooth any rough or sharp edges where molten metal will flow.

# Drying the Casting Area

Torch the casting area for about a minute, this will dry out excess moisture.

# Drying the Casting Area

Replace cope (top section)

# Filling the Crucible

Fill crucible with scrap metal, place the crucible in the furnace and ignite the mixture (Air + Propane).
Feed metal into the exhaust hole as fast as it will take it until the crucible is full.

# How long do I melt?

**After the last piece of aluminum has turned molten** (liquid), let the furnace run for 3 more minutes to reach pouring temperature. With brass, 5 minutes. Bronze, 10 minutes.

# Removing the Dross (Slag)

Turn off the furnace, remove the lid, scoop off the dross (slag) that is floating on top. Remove dross with a slotted steel spoon, a cheap kitchen one will do.

# Pouring the Fluid

While pouring, place a brick on or clamp each mold to prevent leaking!
The aluminum pouring temp is 1325 to 1400F. Brass about 1950F. Bronze 2250F.

# Aluminum Foundry

The excess aluminum is recycled in ingots.

# Shake-out mold into sand Container

Remove the foundry piece and cut off the sprue and vent.

# Remove the Sprue

Cut the sprue and remove the excess sand before the machining process.

# Forging of Metals
## Metallurgical Process

## Forged Aluminum

Forging is a manufacturing process where metal is pressed, pounded, or squeezed under great pressure to produce high-strength parts.

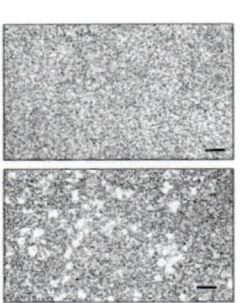

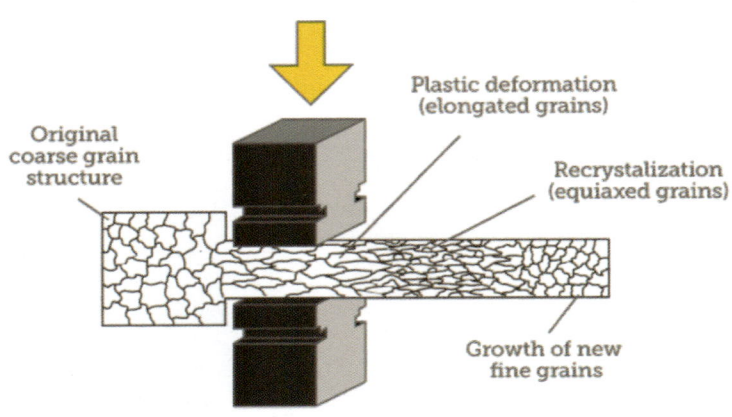

How the open die forging process affects the crystal structure.

## Open-die Forging

Aluminum blocks weighing up to 200,000 pounds and 80 feet in length can be open-die forged to create large aluminum components with optimal structural integrity.

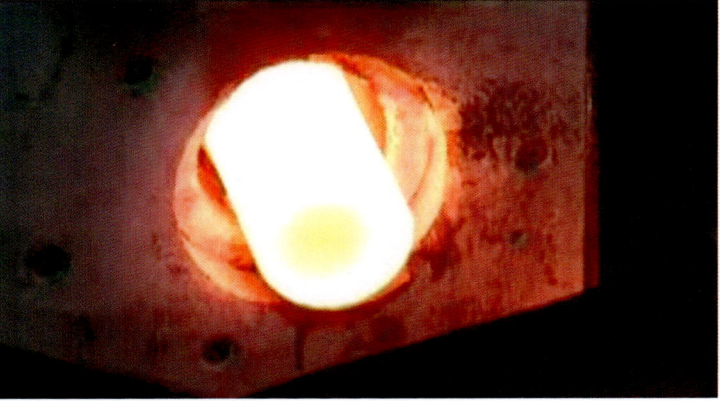

# Close-die Forging

As the name implies, two or more dies containing impressions are brought together as forging stock undergoes plastic deformation. Because the dies restrict metal flow, this process can yield more complex shapes and closer tolerances than open-die forging.

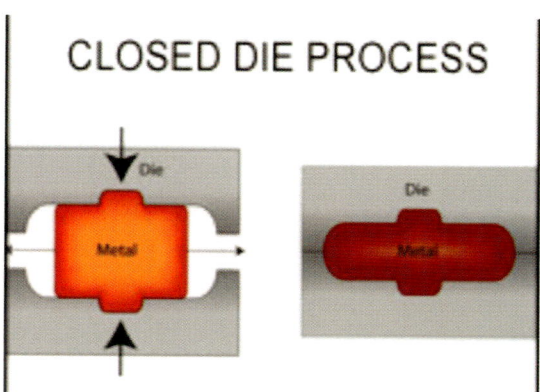

# Hot Extrusion

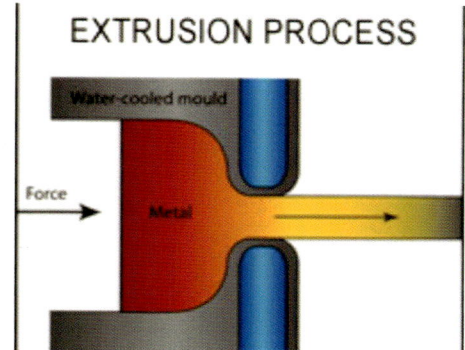

In extrusion, a bar or metal is forced from an enclosed cavity via a die orifice by a compressive force applied by a ram.

Extrusion products include rods and tubes with varying degrees of complexity in cross-section.

# Cold Drawn

Drawing is the pulling of a metal piece through a die by means of a tensile force applied to the exit side. A reduction in cross-sectional area results, with a corresponding increase in length.

# Cold Drawn Advantages

Metals can be formed to much closer dimensions by drawing than by rolling.

Drawn products include wires, rods, and tubing products.

Cold drawn material offer extensive advantages, among other things:

- Smooth and scale-free surface
- Sharp edges
- Uninterrupted grain orientation
- Increased tensile strength and yield point

# Hot Rolling

Rolling is the most widely used deformation process. It consists of passing metal between two rollers, which exert compressive stresses, reducing the metal thickness.

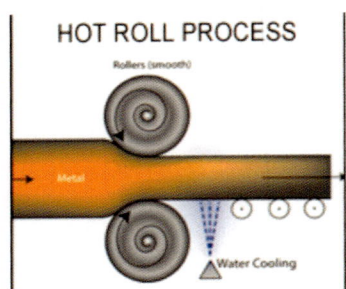

 The initial breakdown of an ingot or a continuously cast slab is achieved by hot rolling. Mechanical strength is improved and porosity is reduced.

## Hot Rolling (Liquid Injected)

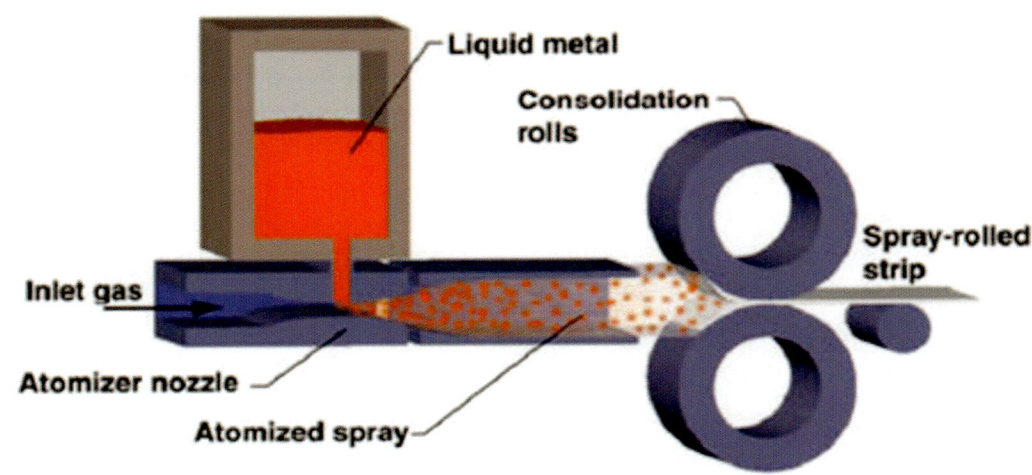

# Forged Aluminum

The forged aluminum wheels on Daytona racecars are a perfect example.